QGIS and Applications in Territorial Planning

QGIS in Remote Sensing Set

coordinated by
André Mariotti

Volume 3

QGIS and Applications in Territorial Planning

Edited by

Nicolas Baghdadi
Clément Mallet
Mehrez Zribi

iSTE

WILEY

First published 2018 in Great Britain and the United States by ISTE Ltd and John Wiley & Sons, Inc.

ISTE Ltd
27-37 St George's Road
London SW19 4EU
UK

www.iste.co.uk

John Wiley & Sons, Inc.
111 River Street
Hoboken, NJ 07030
USA

www.wiley.com

Library of Congress Control Number: 2017961026

British Library Cataloguing-in-Publication Data
A CIP record for this book is available from the British Library
ISBN 978-1-78630-189-5

Contents

Introduction

Scientific and technical issues related to territorial planning have long been firmly accompanied by the software for the graphical representation of the results of these studies. The software is a key element for communication both upstream and downstream of a project, for the phases of design, consultation or conciliation with public and private actors in connection with the infrastructures in question. In this way, Geographic Information Systems (GISs) have, over the last 15 years, taken a central place for all applications and issues related to territorial planning. With a growing number of tools, stakeholders (geomaticians, users, decision makers, scientists), data and references available, we have moved from a mainly cartographic use to more and more advanced analysis, simulation and diagnostics. All spatial scales are concerned, from the very small to the very large, from land use on a European scale to local city planning maps.

This third volume is devoted to the presentation and implementation in QGIS (Quantum Geographic Information System) and its libraries of several applications in relation to territorial planning. The proposed chapters illustrate the great diversity of study cases, spatial scales and actors involved in this problem: from the global scale to the city level, from tailored specific object detection to more general infrastructure location studies, in urban and agroforestry environments, without neglecting the coastal areas.

This work, carried out by scientists of a high technical level, is addressed to geomatics research teams, second-level students (engineering schools, master's) and postgraduate studies (PhD students), and engineers involved in the management of water and territory resources. In addition to the texts of the proposed chapters, readers will have access to the data and tools, as well as screenshots of all the QGIS windows, which illustrate the manipulations necessary to carry out each step of each application. Through this educational work, we wish to contribute to the appropriation of the free software GIS, in connection with remote sensing applications.

A supplement to the chapters, including datasets and screenshots illustrating the practical application of the chapters, is available at the following address:

Using Internet Explorer: ftp://193.49.41.230
Using a FileZilla client: 193.49.41.230

Username: **vol3_en**
Password: **34voL@en3**

We would like to thank everybody who contributed to the development of this book; first of all, the scientists, authors of the chapters of course, but also the experts of the scientific committee for their inspection of the chapters and the corrections made. This project was carried out thanks to the support of IRSTEA (French Research Institute of Science and Technology for Environment and Agriculture), IGN (French National Institute for Geographic and Forestry Information), CNRS (French National Center for Scientific Research) and CNES (French National Center for Space Studies).

We are very grateful to Airbus Defense and Space, CNES and Equipex Geosud for providing us with SPOT-5/6/7 images. Please note that these images may only be used in a research and training framework and any commercial activity based on the data provided is strictly prohibited.

Our thanks also go to our families for their support and to André Mariotti (Professor Emeritus, Pierre and Marie Curie University) and Pierrick Givone (President, IRSTEA) for their encouragement.

Nicolas BAGHDADI
Clément MALLET
Mehrez ZRIBI

1

Design and Implementation of Automated Atlas

1.1. From map to atlas

Given the multiplication of spatial data, automated workflows are becoming important in GIS environments (graphical modeler, programming). Beyond data processing (geometry, topology, attributes, analysis), automation can also be used for symbology or layout to facilitate map design and publication.

If your organization publishes printed or online maps, you often would need to create many maps with the same template – usually one for each administrative unit or region. With the evolution of print composers, GIS software offer more and more layout features, notably for the creation of atlases enabling the compilation of ordered maps and information in the template. Unlike isolated maps, the atlas provides readers with a more in-depth representation of the spaces and themes addressed by combining maps at different scales and graphic or text elements.

The automation of maps and indicators in the form of standardized templates allows both time saving for the production of maps but also a greater graphical consistency of the maps by the homogenization of the layout. This dimension is particularly interesting for the setting up of a graphic charter in the publishing of cartographic documents. This chapter proposes to explore several methods and tools to produce new indicators and to implement a homogeneous, original and stylized cartographic atlas with the QGIS software.

Chapter written by Boris MERICSKAY.

1.2. Automation of maps and indicators

The objective of this chapter is to automate with QGIS the production of an atlas, combining maps and indicators, of the Corsica local region. By using different datasets at the municipal level (population census, agricultural census) and by using several spatial analysis tools, the idea is to initially calculate several key indicators aiming at characterizing these new territories. The combination of different maps and key figures provides the reader with additional complementary elements (statistics and maps).

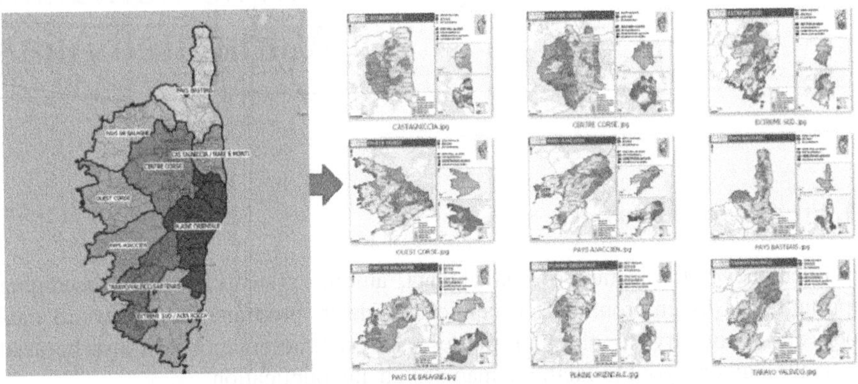

Figure 1.1. *Atlas of Corsica local regions. For a color version of the figure, see www.iste.co.uk/baghdadi/qgis3.zip*

The implementation of this atlas is based on automated map production combined with several indicator-creation processes. Figure 1.1 illustrates the processing steps for atlas implementation in QGIS. To facilitate its reading, processing and handlings are grouped into five main stages, from the conceptualization to the publication of the atlas:

1) atlas template designing;

2) data preparation and indicators creation;

3) atlas implementation in QGIS environment;

4) atlas implementation in the print composer;

5) atlas publication.

2

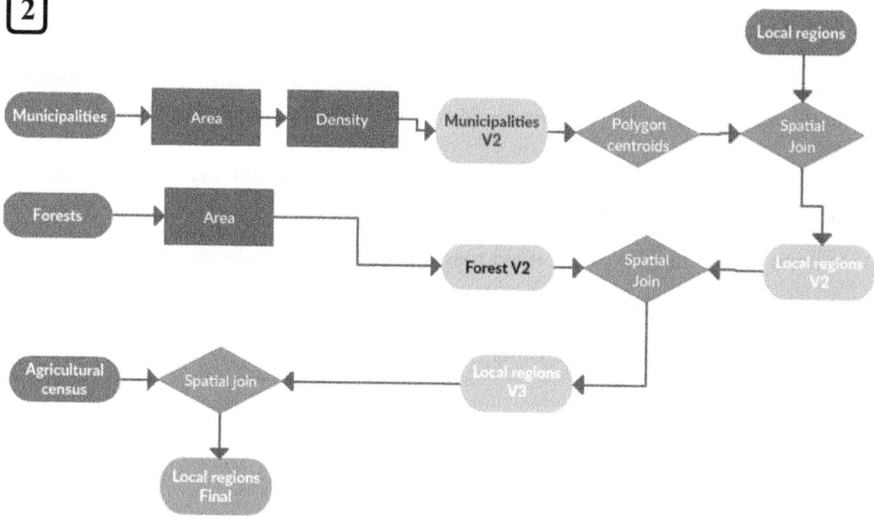

Figure 1.2. *Workflow for calculating indicators*

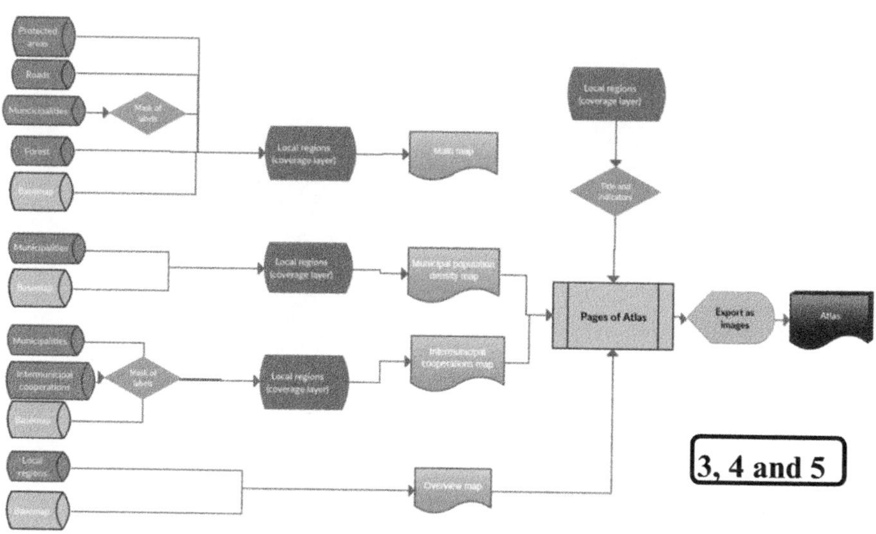

Figure 1.3. *Handling flow for atlas implementation*

1.2.1. *Step 1: atlas template designing*

The first step is to design the atlas template (items and layout). The items depend on the objective of the atlas (communication, decision-making and analysis).

We propose in this chapter an atlas template to present the local regions of Corsica based on seven items as illustrated in Figure 1.4:

1) main map (municipalities, main roads, protected areas, forests);

2) municipal population density map;

3) intermunicipal cooperations map;

4) overview map;

5) name of local region;

6) logo;

7) indicators.

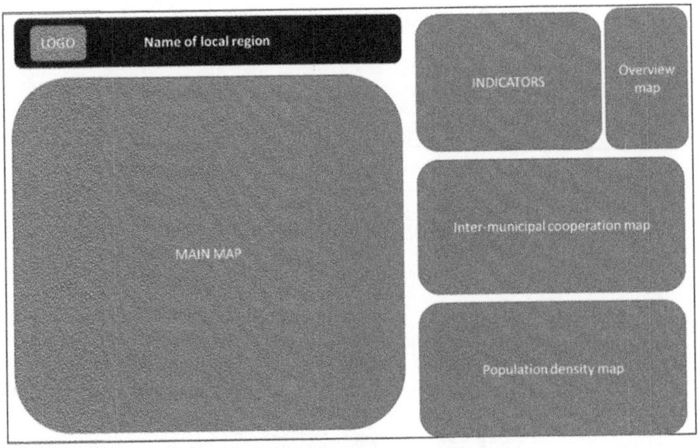

Figure 1.4. *Atlas template*

1.2.2. *Step 2: data preparation and indicators creation*

The second step is to prepare the datasets (reproject, transformation) and create indicators for the map atlas. The objective of this step is to enrich the atlas coverage layer of basic and derived statistics (number of communes, population, forest area, number of farms, etc.) to create indicators.

Datasets use different coordinate reference systems (CRS) – WGS 84 and Lambert 93. To facilitate processing it is necessary to harmonize the CRS of spatial datasets using the Lambert 93 projection as the reference coordinate reference system for all layers of the project.

QGIS functionalities:
- Reproject layer: *QGIS geoalgorithms > Vector general tools*

1.2.2.1. *Calculate basic statistics*

***Municipalities* layer**: Calculate the municipal population density.

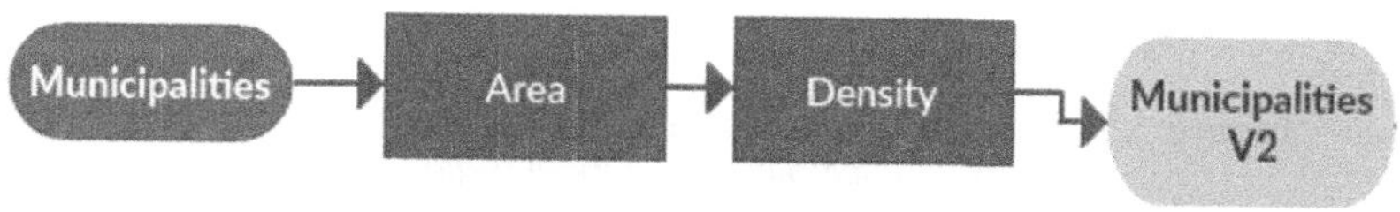

***Forests* layer**: Calculate the area of forests.

QGIS functionalities:
- Field Calculator

1.2.2.2. *Aggregating municipal data at the scale of local regions*

The second step of data preparation is to change the analysis scale by aggregating municipal data at the local regional scale. In order to carry out this transformation of the data, it is necessary to mobilize the **spatial join**[1], which makes it possible to associate and aggregate attributes from one layer to another according to topological relationships between objects in space (intersect, within, contains). A spatial join involves matching rows from the Join Features to the Target Features based on their relative spatial locations.

To spatially join the attributes of the common layer to that of the project territories, it is important to be vigilant about the topology of the objects. Indeed the topological consistency that defines the quality of the join is not always good. As illustrated in Figure 1.5, the topological coherence of the boundaries of municipalities and project boundaries is not of good geomatric quality.

---

1 Joins attributes from one feature to another based on the spatial relationship. The target features and the joined attributes from the join features are written to the output feature class.

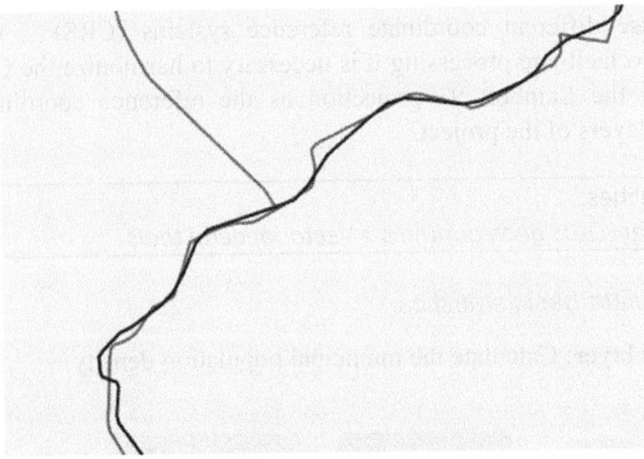

Figure 1.5. *Example of topological errors between the boundaries of municipalities and local regions*

This current situation can be solved by transforming the geometry of one of the layers and more precisely by using the polygonal centroids to facilitate spatial joining (Figure 1.6).

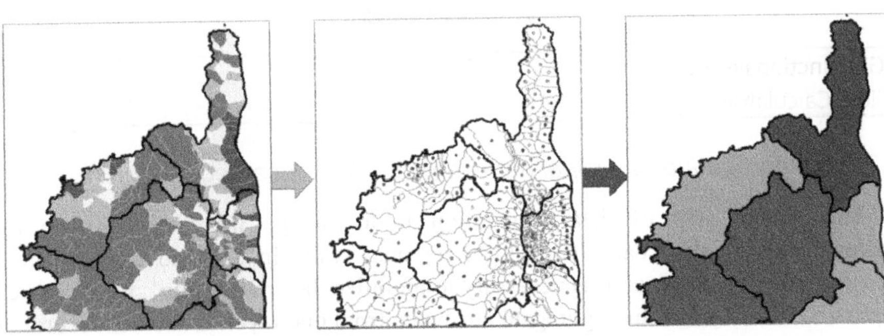

Figure 1.6. *Polygons spatial join based on centroids. For a color version of the figure, see www.iste.co.uk/baghdadi/qgis3.zip*

We must also be vigilant about the geometric transformation of polygons in points. As shown in Figure 1.7, the centroid of the municipality of Saint-Florent is located outside the perimeters of local areas. It is therefore necessary to replace the centroid within the boundaries of the project territories in order not to compromise the spatial junction.

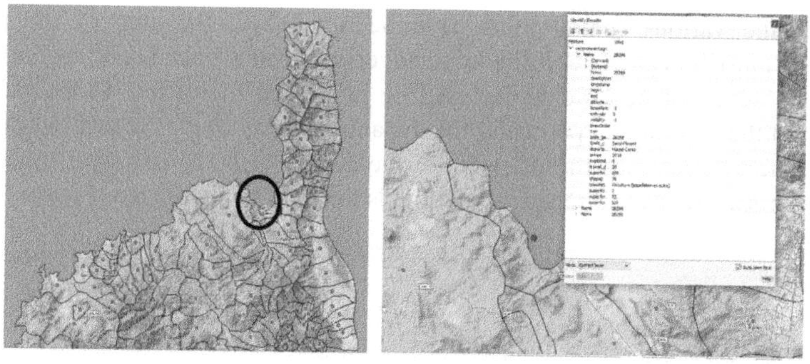

Figure 1.7. *Example of topological error following the transformation*

> **QGIS functionalities:**
> - Polygon centroids: *QGIS geoalgorithms > Vector geometry tools OR SAGA > Vector polygon tools*
> - Move feature: *Digitizing toolbar*

Aggregating municipal population at the scale of local regions

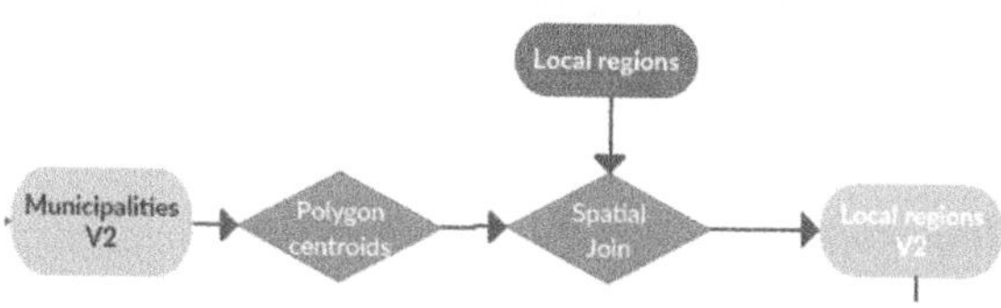

Aggregating forest areas at the scale of local regions

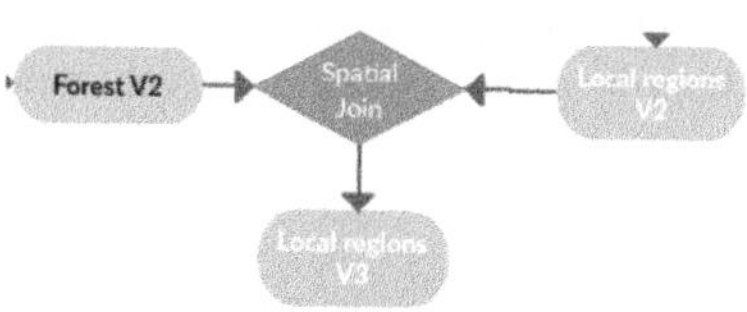

Aggregating agricultural census data at the scale of local regions

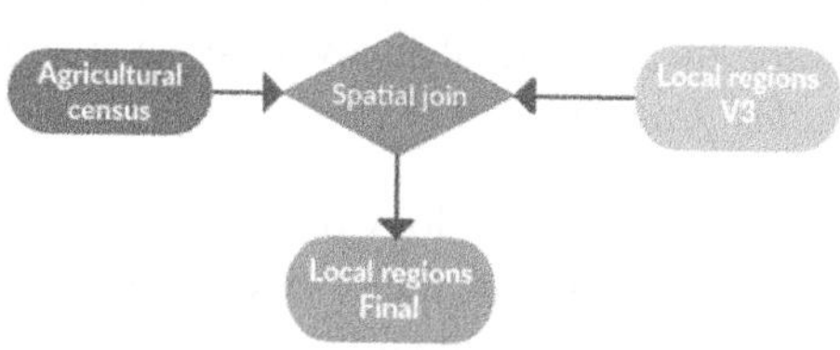

Once the various spatial joins have been realized, the area of the project territories must be calculated. The atlas coverage layer (*Final Local regions*) now contains a series of new fields that will be mobilized thereafter. To facilitate the creation of the atlas and in particular the dynamic display of the created indicators, it is advisable to clean the attribute table of the *Final Local regions* layer by deleting the unnecessary fields and renaming the fields that will be mobilized (Figure 1.8)

	objectid 1	pays	Nb commune	Population	Surface TP	Sup foret	Nb exploit	Sup Agrico	Cheptel
1	1.00000	PAYS DE BALAGNE	36	22823	957.38	309	345	21395	13092
2	2.00000	CASTAGNICCIA / MARE E MONTI	59	20451	378.87	19	320	10036	6547
3	3.00000	CENTRE CORSE	56	16306	1347.73	422	347	32403	18519
4	4.00000	EXTREME SUD / ALTA ROCCA	21	25439	1065.14	263	247	12799	6672
5	5.00000	OUEST CORSE	33	7596	923.22	248	189	18318	8304
6	6.00000	PAYS AJACCIEN	25	98273	859.31	207	256	12358	6910
7	7.00000	PAYS BASTIAIS	53	93124	967.04	13	342	18588	10395
8	8.00000	PLAINE ORIENTALE	34	20791	1224.57	302	471	23325	12047
9	9.00000	TARAVO/VALINCO/SARTENAIS	43	15405	1036.49	264	293	18682	9214

Figure 1.8. *Attribute table of Final Local regions layer after attribute table cleaning*

QGIS functionalities:
- Spatial Join: *QGIS geoalgorithms > Vector general tools > Join attributes by location*
- Attribute table cleaning: *Fields calculator > Erase field*

1.2.3. *Step 3: atlas implementation in QGIS project*

Once the data have been prepared, it is time to move on to formatting. The idea is to provide here a series of tips to optimize the formatting of the data in order to produce aesthetically pleasing, comprehensible and original maps that will be integrated into the atlas boards (atlas coverage layer, rule-based display, masks, custom labels).

1.2.3.1. *Configure atlas coverage layer*

The first step is to configure the atlas coverage layer (Figure 1.9). It is on the objects of this layer that the atlas will be based for map automation. To define this parameter, it is necessary to apply to the coverage layer a rule-based style by mobilizing the filter *@atlas_featureid = $ id.*

It is also possible to mobilize a symbology of the Inverted Polygons type so that the emphasis at the graphic level is put on the selected project territory. They need to be rule based (style) and need to contain the following rules (in Style):

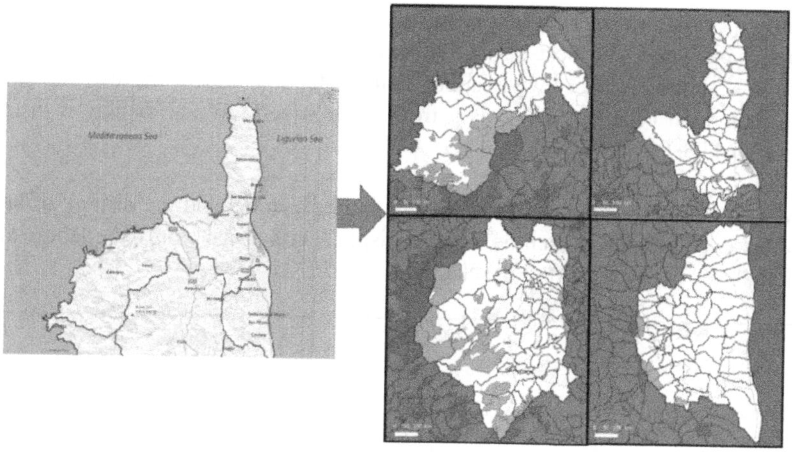

Figure 1.9. *Principle of atlas coverage layer. For a color version of the figure, see www.iste.co.uk/baghdadi/qgis3.zip*

QGIS functionalities:
- Rule-based style: *Layer properties > Style > Rule*
- Inverted polygons style: *Layer properties > Style > Inverted polygons*

1.2.3.2. *Designing the main map*

The main map represents the local regions by combining reference data and environmental data relating to protection:

– local regions;

– municipalities with labels (names);

– main roads;

– forests;

– protected environmental areas (ZNIEFF);

– basemap.

The idea here is to combine several modes of representation of the data in order to obtain aesthetically pleasing, clear and customizable maps that will be generated in an automated way (atlas filter, inverted polygon and label mask).

The choice of the basemap is an important step for map design. With Web services as WMS (Web Map Service), it is possible to integrate a diverse selection of basemaps into QGIS. In addition to the classical basemaps like streets or imagery, it is now possible to mobilize more minimalist or customized basemaps for use as adapted basemaps. The best service to personalize and use original basemaps in QGIS is Mapbox (https://www.mapbox.com). Mapbox allows the user to quickly create their own basemap using data from OpenStreetMap and display it in QGIS using a WMTS.

After configuring the layout of the layers, the basemap and the display of labels, the Mask extension must then be used to display with a mask only the labels of the municipalities of the local region selected.

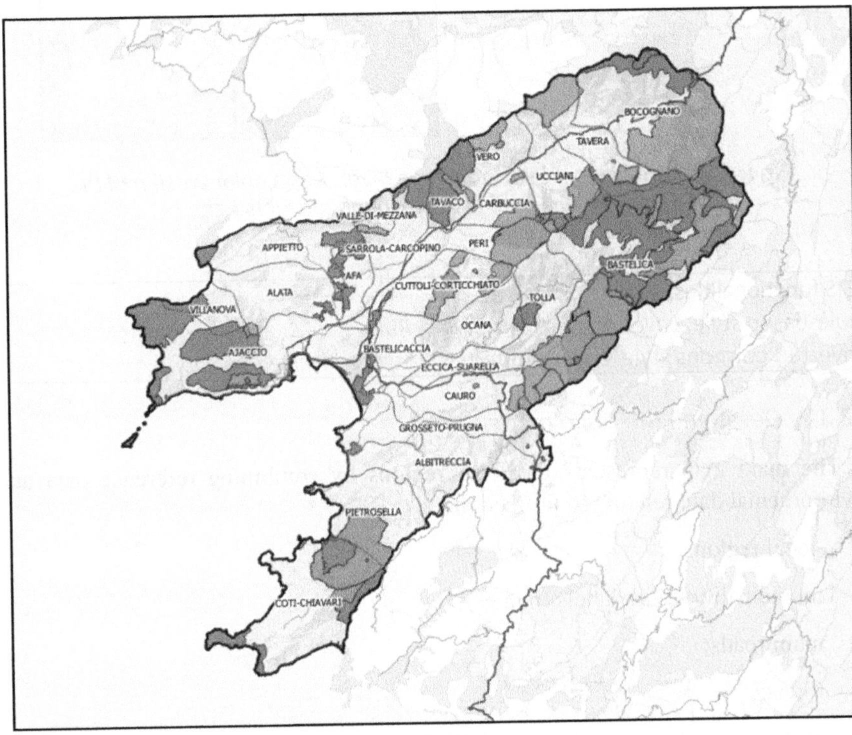

Figure 1.10. *Example of layout of main map. For a color version of the figure, see www.iste.co.uk/baghdadi/qgis3.zip*

QGIS functions:

- Symbology: *Layer properties > Style*
- OpenLayers Plugin or QuickMaps Services to add basemaps from Internet
- *Mask* Extension to display only certain labels
- Inverted Polygons: *Layer properties > Style*
- Original basemap: WM(T)S *connection*

1.2.3.3. *Municipal population density map*

The objective is here to design a map to present the population density of the municipalities (Figure 1.11). It is necessary to duplicate the layer in the layer manager in order to apply a symbology in plots of colors to represent the values.

The classification step is particularly important and requires real work and reflection on the part of the map designer. Indeed, the different automated methods of discretization proposed in GIS software must be used with caution, since they are rarely relevant.

A good classification concerns both the choice of the number of classes and the selection of a data classification method. The reflection and labor of the cartographer must mobilize the histogram of the values and the basic statistics of the statistical series to be generalized (mean, standard deviation, minimum and maximum). As is often the case, the map maker uses a type of data classification to produce their own unique choropleth map. Each data classification method impacts the reader differently.

Figure 1.11. *Example of municipal population density map. For a color version of the figure, see www.iste.co.uk/baghdadi/qgis3.zip*

QGIS functionalities:
- Symbology: *Layer properties > Style > Graduated*
- Classification: *Histogram of Distribution, Statist (extension), Basic statistics for numeric fields*

1.2.3.4. *Intermunicipal cooperations map*

The objective is to produce here a map representing the various Corsican Intermunicipal cooperations (Figure 1.12). In addition to applying a categorizated symbology, it is also necessary to configure the label display of the Intermunicipal cooperations names (update the Mask extension to apply a label mask on the Intermunicipal cooperation layer). To complete the map, it is necessary to duplicate the Intermunicipal cooperation layer to apply a suitable formatting (blank contours without filling for example).

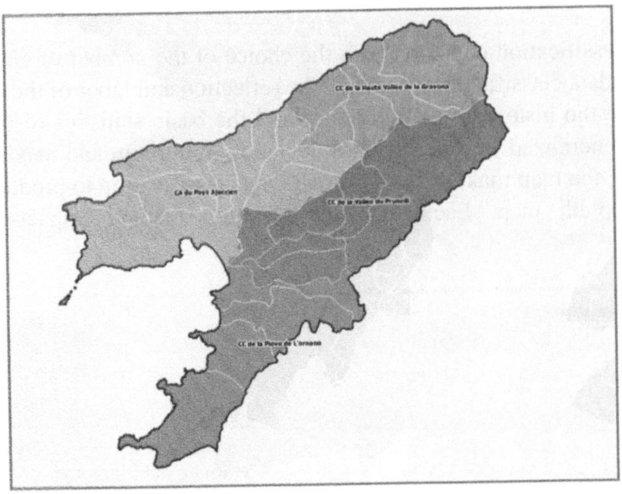

Figure 1.12. *Example of an Intermunicipal cooperation map in Corsica. For a color version of the figure, see www.iste.co.uk/baghdadi/qgis3.zip*

QGIS functions:
- Symbology tool: *Layer properties*
- *Mask* extension to display only labels of the selected project territory

1.2.3.5. *Overview map*

The objective here is to produce an overview map which will allow us to locate the project territory of the atlas plate at the level of Corsica. It is necessary to duplicate the layer of the project territories and to apply to it a simple symbology.

1.2.4. *Step 4: atlas implementation in print composer*

1.2.4.1. *Configuration and implementation of atlas*

The first step to implement an atlas in print composer is to configure and generate the atlas. To do this, you must activate the atlas functions, define the coverage layer and configure the main map (dimension, position, frame, scale, legend, arrow, sources). Once the map is implemented, you must lock the layer and its style in the properties of the map (Figure 1.13).

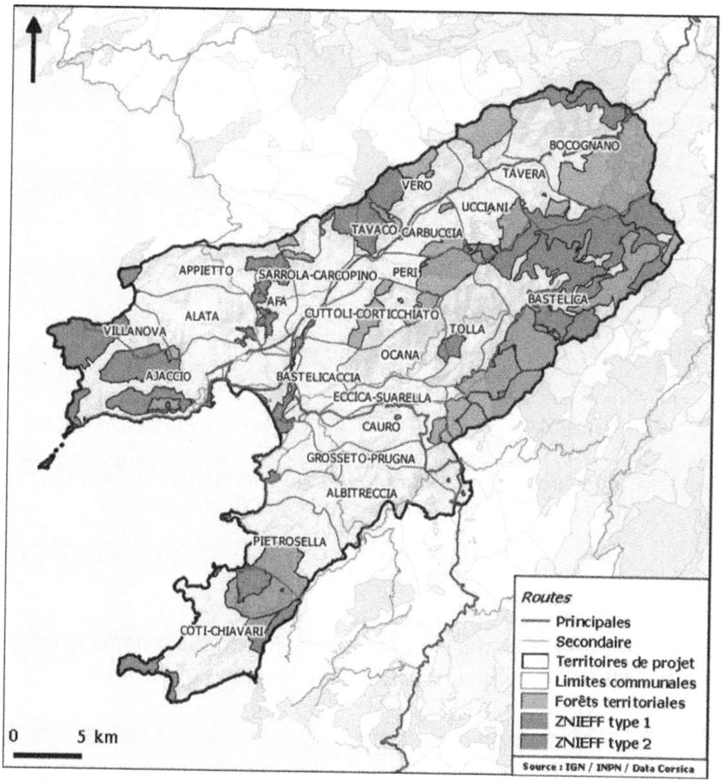

Figure 1.13. *Add main map to the atlas template. For a color version of the figure, see www.iste.co.uk/baghdadi/qgis3.zip*

1.2.4.2. *Add dynamic items (title, indicators)*

In this step, it is necessary to add to the pages of the atlas a series of dynamic items, which will update according to the pages of the atlas:

– title (based on the field "pays" of the *local region* layer);

– indicators (based on the fields computed and aggregated upstream) with the mobilization of pictograms for an infographic-type rendering.

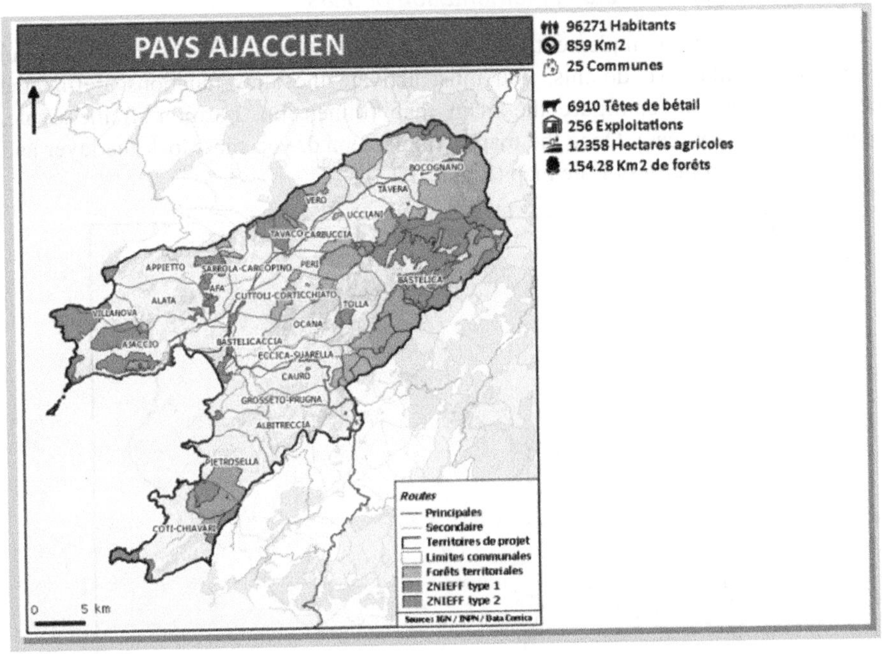

Figure 1.14. *Add title and indicators to the atlas template. For a color version of the figure, see www.iste.co.uk/baghdadi/qgis3.zip*

1.2.4.3. *Add static items*

The idea here is to enrich the pages of the atlas with static items (such as logo or pictos) that will always be similar regardless of the pages of the atlas.

1.2.4.4. *Add municipal population density*

The aim here is to activate a symbology for the municipal population density map in the QGIS project. Then add the map legend, scale and source. It is also necessary to configure the parameters controlled by the atlas. Once the map is implemented, it should lock the layer and its style in the properties of the map.

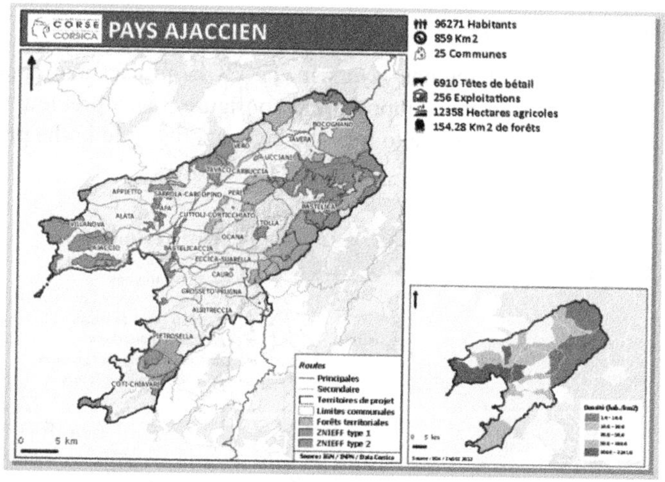

Figure 1.15. *Add municipal population density map to the atlas template. For a color version of the figure, see www.iste.co.uk/baghdadi/qgis3.zip*

1.2.4.5. *Add intermunicipal cooperation map*

As before, the aim here is to activate the symbology of the Intermunicipal cooperation layers in the QGIS project. Next, configure the map in print composer and add the scale and source. Once the map is implemented, it should lock the layer and its style in the properties of the map.

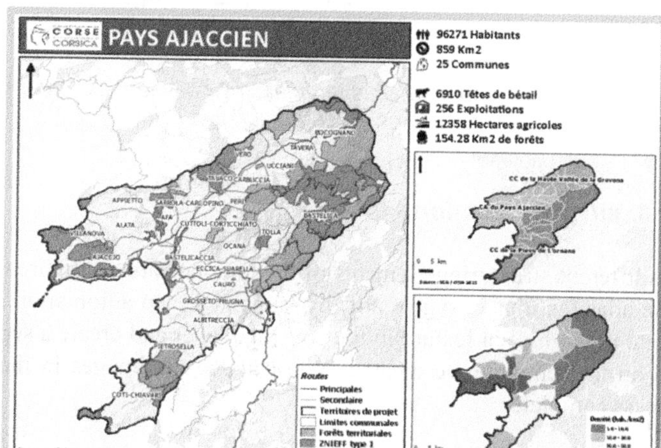

Figure 1.16. *Add Intermunicipal cooperation map to the atlas template. For a color version of the figure, see www.iste.co.uk/baghdadi/qgis3.zip*

1.2.4.6. *Add overview map*

First, activate the Local regions layers in the QGIS project and apply a basic symbology. Add the map to the composer and configure it as a preview map. Once the map is formatted, it is advisable to lock the layer and its style in the properties of the map.

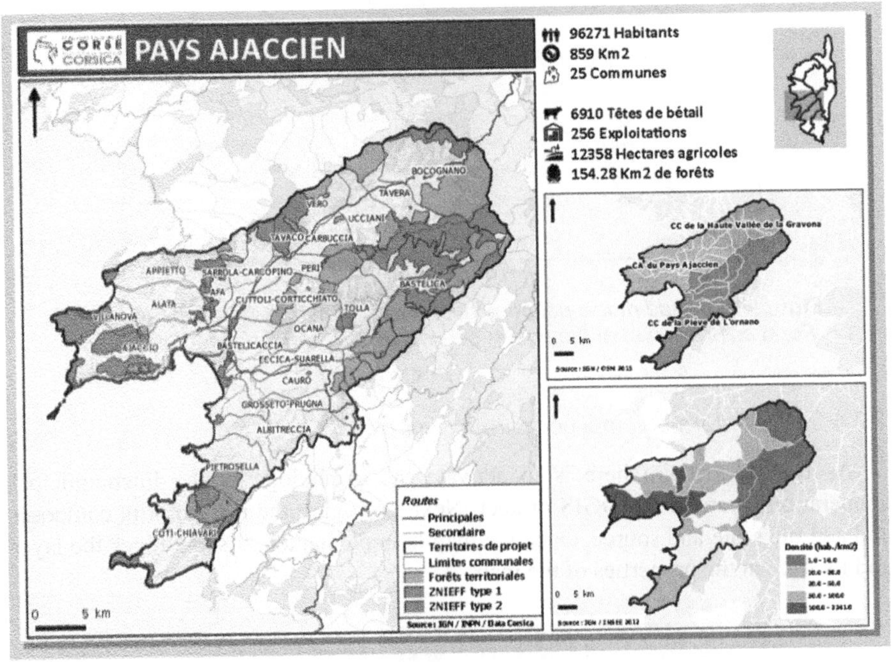

Figure 1.17. *Example of completed atlas template. For a color version of the figure, see www.iste.co.uk/baghdadi/qgis3.zip*

1.2.5. *Step 5: atlas publication*

Once the different structuring elements of the template are configured, it is time to publish the atlas (as image, svg or pdf) by launching map automation. The Atlas tool will iterate through each feature in the coverage layer and create a separate map image based on the template we created. You can see the images in the directory once the process completes.

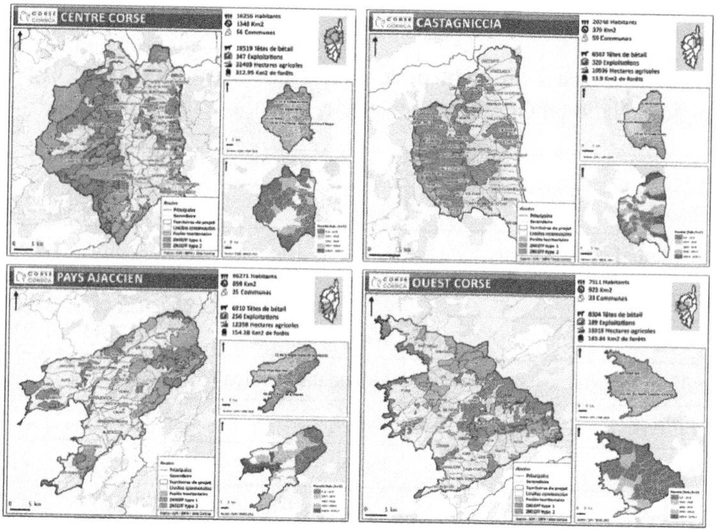

Figure 1.18. *Overview of the pages of the atlas. For a color version of the figure, see www.iste.co.uk/baghdadi/qgis3.zip*

1.3. Implementation of the application

1.3.1. *Software and data*

1.3.1.1. *Software required*

The process discussed in this chapter uses the basic functionality of the QGIS software (version 2.18). It is also necessary to install some additional extensions to carry out the processing chain: GroupStats, Mask, OpenLayerPlugin, Statist.

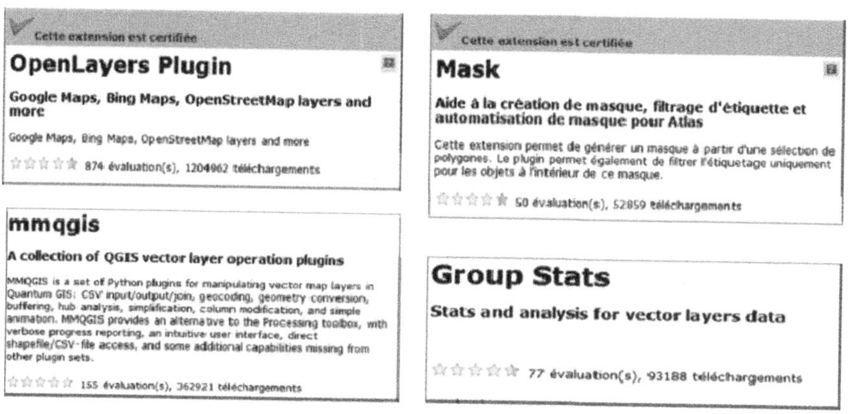

Figure 1.19. *QGIS extensions*

1.3.1.2. *Data*

This chapter is based on eight open data datasets from different organizations (Corsica region, National Geographic Institute, OpenStreetMap).

Name layer	Source	Year	CRS
Municipalities	IGN	2016	2154
Intermunicipal cooperations	OSM	2016	4326
Local regions	Opendata Corsica	2017	4326
Main roads	IGN	2017	2154
Agricultural census	Opendata Corsica	2010	4326
ZNIEFF type 1	INPN	2016	2154
ZNIEFF type 2	INPN	2016	2154
Forest	Opendata Corsica	2017	4326

Table 1.1. *Summary of the data used*

Here are some links for downloading referential data:

1) **Municipalities**

Name of layer: GEOFLA® 2016 v2.2 Communes France Métropolitaine

URL: http://professionnels.ign.fr/geofla

2) **Intermunicipal cooperation**

Name of layer: Contours des EPCI 2015 (simplifié à 100 m)

URL: https://www.data.gouv.fr/fr/datasets/contours-des-epci-2015

3) **Local regions**

Name of layer: Territoires de projet de la Collectivité Territoriale de Corse

URL: https://www.data.corsica/explore/dataset/territoires-de-projet-de-la-collectivite-territoriale-de-corse/export

4) **Main roads**

Name of layer: ROUTE 120®

URL: http://professionnels.ign.fr/route120

And here are the links for downloading thematic data:

5) Agricultural census

Name of layer: Recensements Agricoles par communes (1988, 2000 and 2010)

URL: https://www.data.corsica/explore/dataset/recensementagricole/export

6) ZNIEFF from type 1

Name of layer : Zones Naturelles d'Intérêt Ecologique Faunistique et Floristique de type 1

URL : https://inpn.mnhn.fr/telechargement/cartes-et-information-geographique/inv/znieff1

7) ZNIEFF from type 2

Name of layer: Zones Naturelles d'Intérêt Ecologique Faunistique et Floristique de type

URL: https://inpn.mnhn.fr/telechargement/cartes-et-information-geographique/inv/znieff2

8) Forests

Name of layer: Forêts de Corse

URL: https://www.data.corsica/explore/dataset/forets-de-corse/export

1.3.2. *Step 2: data preparation and indicators creation*

1.3.2.1. *Calculate municipal population density*

Process	Handling in QGIS
1. Select Corsica municipalities	In **QGIS**: • Open Attributes table  In Select by expression: • Write the query: "NOM_REG" = 'CORSE' In the Attributes table: • Invert the selection • Switch to Editing mode • Delete selected features • Stop Editing mode and save

2. Calculate municipal population density	In **Field calculator**: • Create new Field "AREA" (decimal type) > Calculate area of municipalities with function *$area* by writing the query "$Area/1000000" (convert m^2 to km^2) In **Field calculator**: • Create new field "Density" (integer type) > Calculate municipalities population density by writing the query "Population/AREA"

Table 1.2. *Steps for the calculation of population density in the layer "Municipalities"*

1.3.2.2. *Calculate forest area*

Process	Handling in QGIS
1. Calculate Forest Area	In **Field calculator**: • Create new Field "AREA" (decimal type) > Calculate Forest Area with the expression *$Area/1000000* (convert m^2 to km^2)

Table 1.3. *Step for the calculation of forest area*

1.3.2.3. *Aggregate municipal data at scale of local regions*

1.3.2.3.1. Aggregate data from municipalities layer at scale of local regions (population and number of municipalities)

Process	Handling in QGIS
1. Create polygon centroids	In **QGIS**: • Vector > Geometry tools > **Polygon centroids** • To facilitate spatial joining, it is recommended to create a new field (integer type) with the value of the column POPULATION > Field Calculator
2. Reproject local region layer	• Save vector layer with CRS 2154 (Lambert 93)

<table>
<tr>
<td>3. Aggregate municipal data at scale of local regions</td>
<td>

- Vector > Data Management Tools > **Join attributes by location**

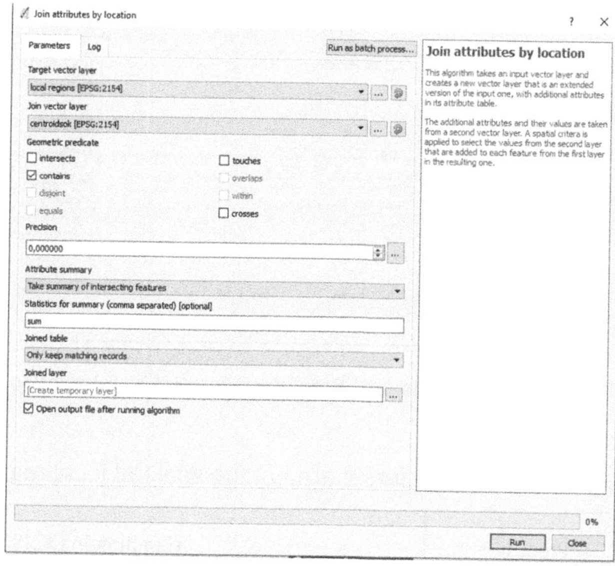

It is essential to verify if the spatial join has performed well at the aggregation level:

- Use a statistical tool to compare the fields of the joined layer and the new fields resulting from the aggregation (Basic statistics for numeric fields or Statist extension)

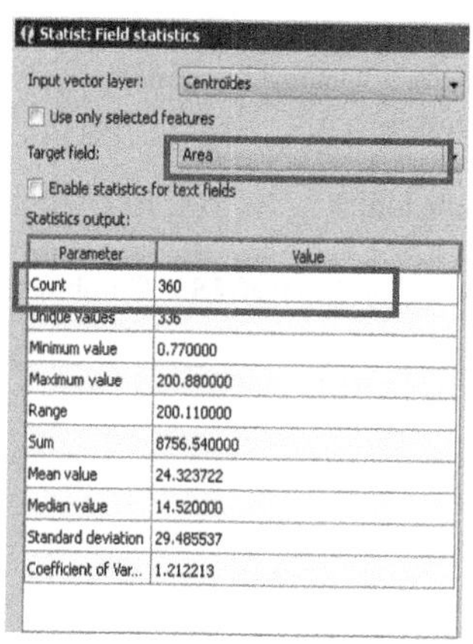

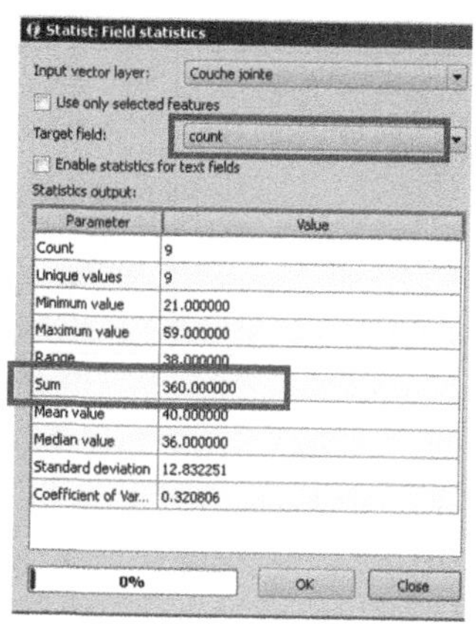

</td>
</tr>
</table>

<table>
<tr><td></td><td>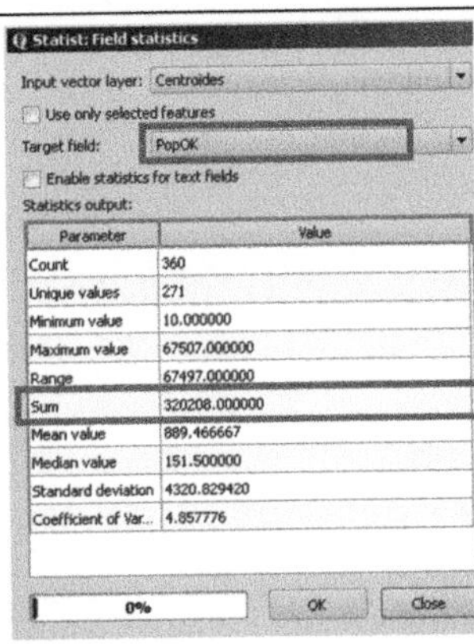

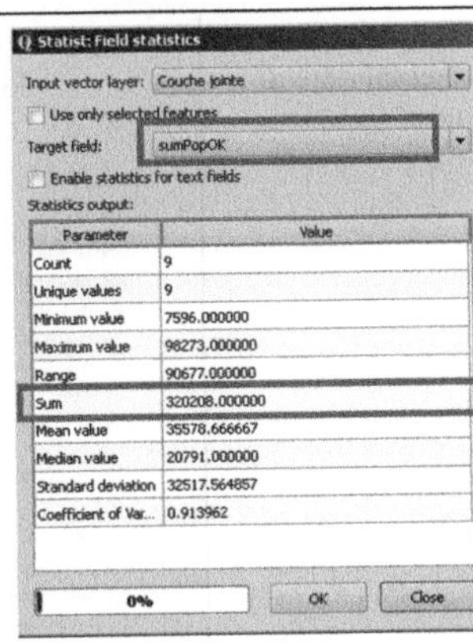
</td></tr>
</table>

Table 1.4. *Steps for aggregating population data*

1.3.2.3.2. Aggregate forest area at the scale of local region

Process	Handling in QGIS
1. Reproject forest layer	In **QGIS**: • Save vector layer with CRS 2154 (Lambert 93)
2. Create polygon centroids	• Vector > Geometry tools > **Polygon centroids**
3. Aggregate data at scale of local regions	• Vector > Data Management Tools > **Join attributes by location** Other available tool: • Extension **MMQGIS** > Combine > Spatial Join

	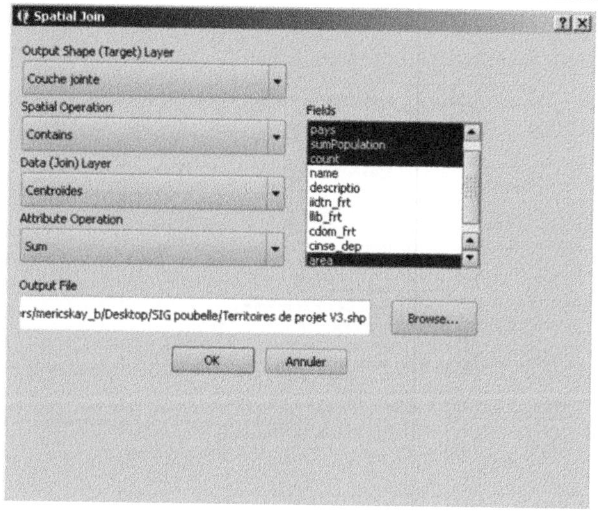

Table 1.5. *Steps for aggregating forest data*

1.3.2.3.3. Aggregate data of the layer agricultural census (livestock/area/number of exploitations)

Process	Handling in QGIS
1. Keep only the 2010 Agricultural census data	In **QGIS**: • Open Attributes table 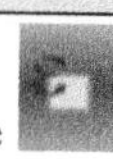 In the attribute selector : • Write the query: "annee" = '2010' In the Attributes table: • Invert selection • Switch to Editing mode • Delete selected features • Stop Editing mode and save
2. Reproject Agricultural census layer	• Save vector layer with CRS 2154 (Lambert 93)
3. Aggregate data at scale of local regions	• Vector > Data Management Tools > **Join attributes by location**

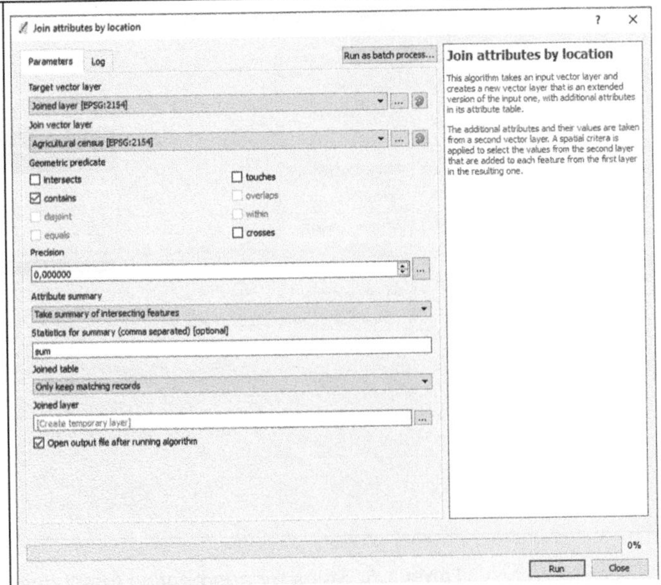

- Check results of the spatial join

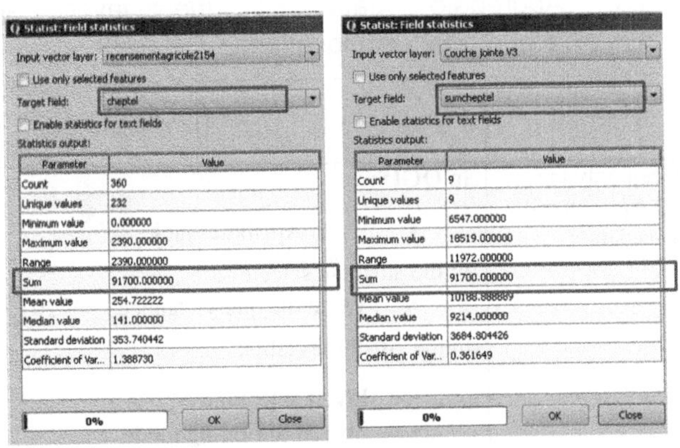

Table 1.6. *Steps for aggregating agricultural census data*

1.3.2.3.4. Calculate area of local region and clean attribute table

1. Calculate area of local region	In Field calculator: • Create new Field "AREA" (decimal type) > Calculate area of municipalities with function *$area* by writing the query *$Area/1000000* (convert m^2 to km^2)
2. Clean Attribute table	In the Attribute table: • Switch to Editing mode • Delete unnecessary fields • Rename fields with field calculator

Table 1.7. *Calculate area of local region and clean attribute table*

1.3.2.4. *Prepare intermunicipal cooperations layer*

Process	Handling in QGIS
1. Reproject intermunicipal cooperations layer	In **QGIS**: • Save vector layer with CRS 2154 (Lambert 93)
2. Keep only the intermunicipal cooperations of Corsica	In **QGIS**: • Vector > Research tools > Select by location • Select intermunicipal cooperations layer that intersects with the municipalities of Corsica In the **Attributes table**: • Invert selection • Switch to Editing mode • Delete selected features • Stop Editing mode and save

Table 1.8. *Steps for the preparation of the intermunicipal cooperations layer*

1.3.3. *Step 3: atlas implementation in QGIS project*

This step consists in shaping the different layers of data that will be mobilized in the various maps of the atlas. The other dimensions of this step revolve around the definition of the style of the coverage layer (on which the atlas is based) and the configuration of the display masks for the labels.

1.3.3.1. *Defining the coverage layer of the atlas*

<table>
<tr><th>Process</th><th>Handling in QGIS</th></tr>
<tr><td>1. Defining the coverage layer of the atlas</td><td>
In the style properties of the coverage layer:
<ul><li>Choose a style based on a set of rules</li></ul>
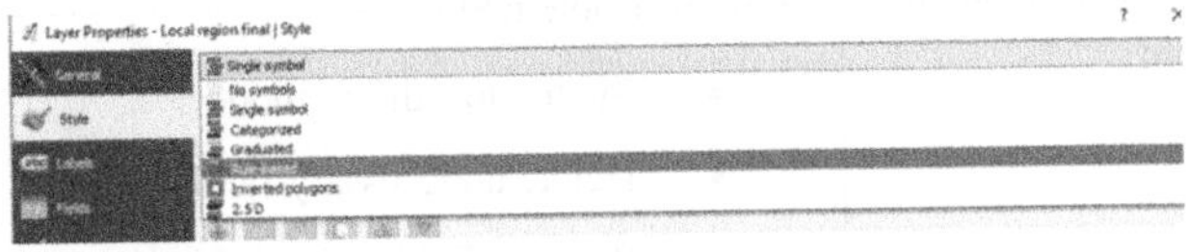

<ul><li>Click on the Rule for edit rule</li>
<li>Click on filter to edit the filter</li></ul>
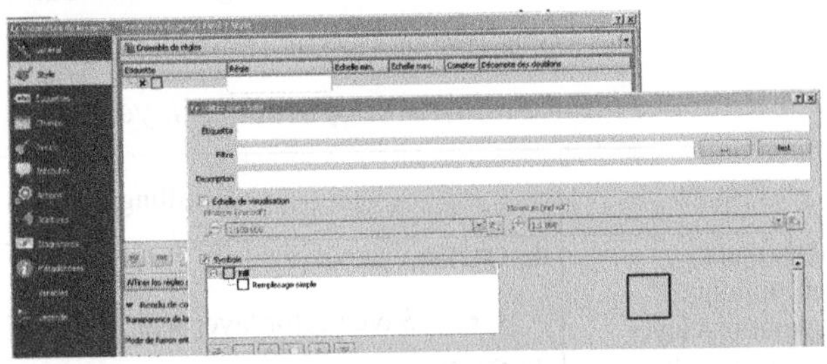
<ul><li>Enter the expression: @atlas_featureid = $id</li></ul>
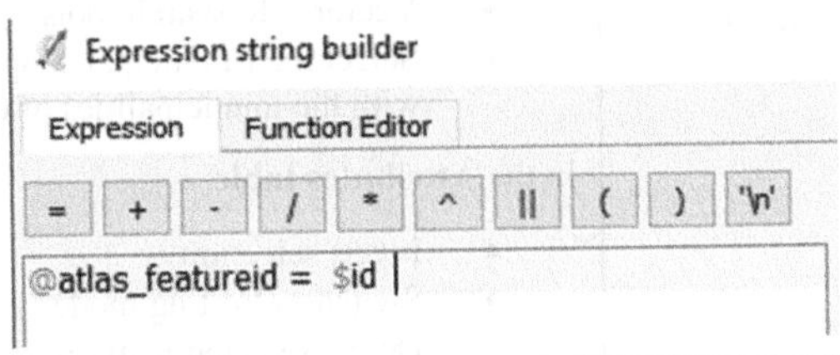

</td></tr>
<tr><td>2. Apply Inverted polygons style</td><td>
In the style properties of the coverage layer:
<ul><li>Select Inverted polygons</li></ul>
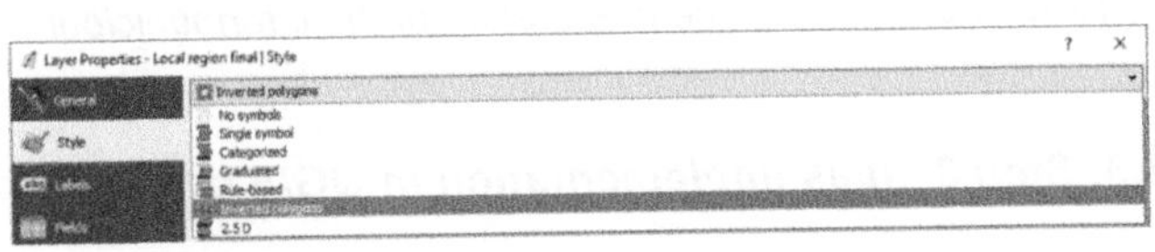

<ul><li>Configure:
<ul><li>white fill</li>
<li>20% of transparency</li>
<li>black outline, width of 1</li></ul></li></ul>
</td></tr>
</table>

	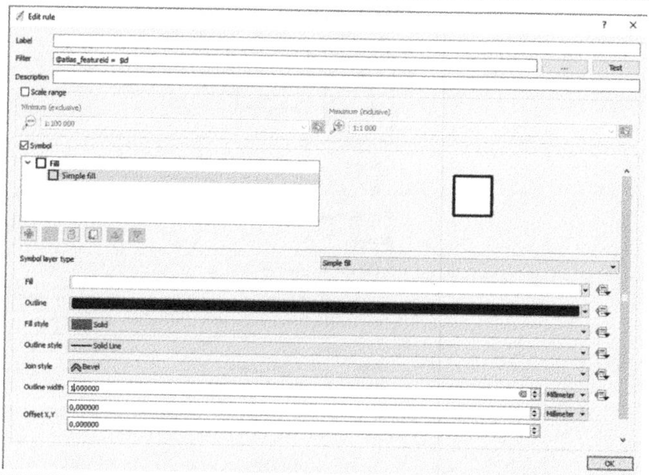

Table 1.9. *Steps for defining the coverage layer of the atlas*

1.3.3.2. *Configuration of municipalities labels mask*

Process	Handling in QGIS
1. Apply a symbology	In **QGIS**: • Format the data • Configure labels of municipalities
2. Apply a mask to display only the municipalities labels of selected feature	In **QGIS**: • **Select a feature of the local region layer**  • **Run the Mask extension** • **Configure the Mask Extension**

	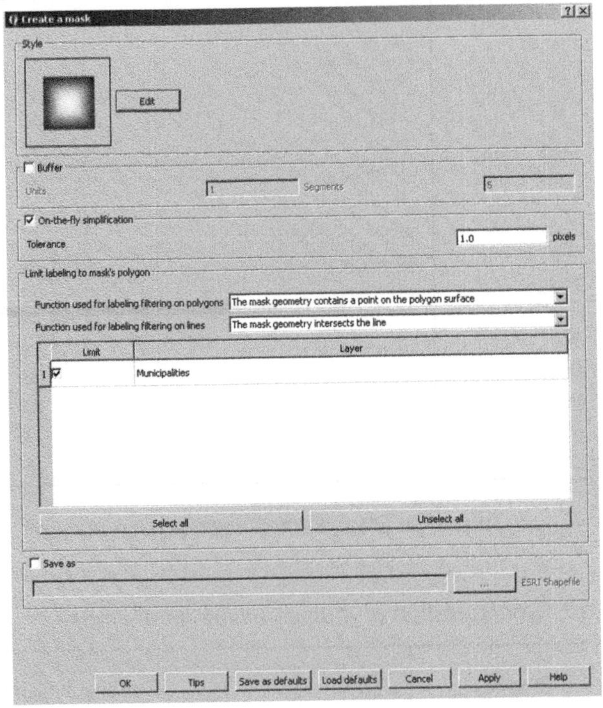

Table 1.10. *Steps for the configuration of municipalities labels mask*

1.3.3.3. *Configuration of mask of intermunicipal cooperations labels*

Process	Handling in QGIS
1. Apply a mask to display only the intermunicipal cooperations labels of the selected local region	In **QGIS**: • **Select a feature of the local regions layer** • **Update the current mask** • **Configure the Mask Extension**

	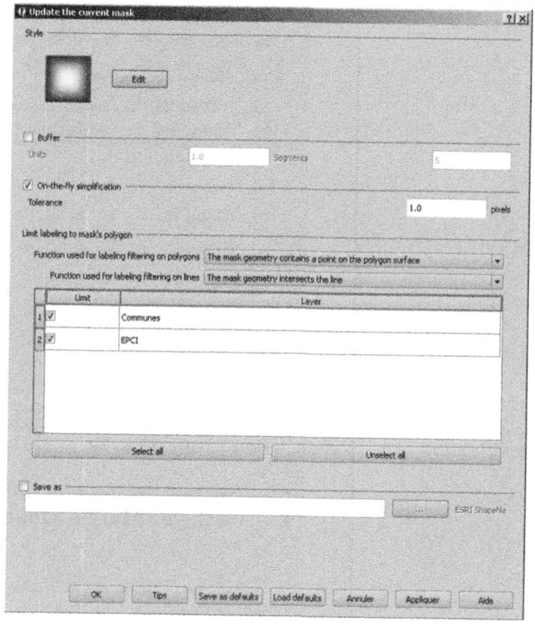

Table 1.11. *Steps for the configuration of intermunicipal cooperations labels mask*

1.3.4. *Step 4: atlas implementation in print composer*

This fourth step has as its objective the configuration and the implementation of the atlas in the composer.

1.3.4.1. *Activation of atlas in print composer*

Process	Handling in QGIS
1. Define map symbology	In **QGIS**: • Activate layers of main map
2. Configuration of coverage layer	In the composer: • Add map  • Activate and configure atlas properties ○ **Click on the Atlas generation tab** ○ **Click on Generate an atlas** ○ **Select the Coverage layer**

<table>
<tr>
<td></td>
<td>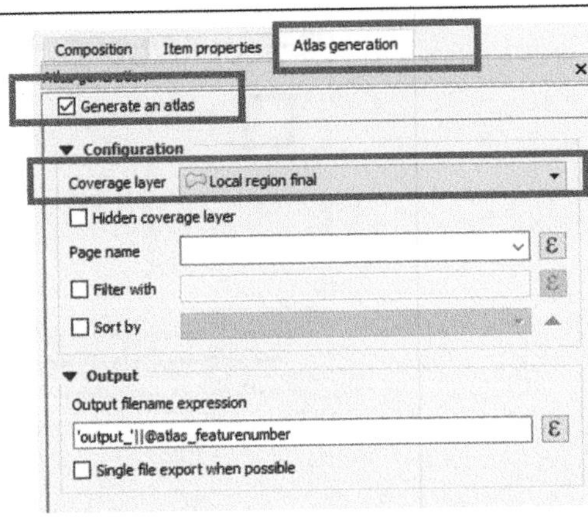
</td>
</tr>
<tr>
<td>3. Configuration of map</td>
<td>In the Item properties tab:
<ul>
<li>Activate the function Controlled by atlas</li>
<li>Configure Margin around feature</li>
</ul>
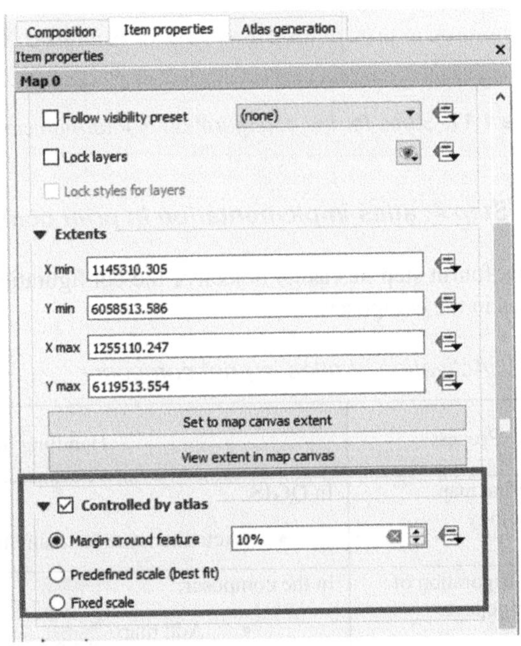
</td>
</tr>
</table>

4. Generate atlas in print composer	In the **composer toolbox**: • Click on Preview atlas • Use the arrows to view the different maps generated automatically: 
5. Lock layer and style layer	In the **Item properties** tab: • Lock layers • Lock styles for layers 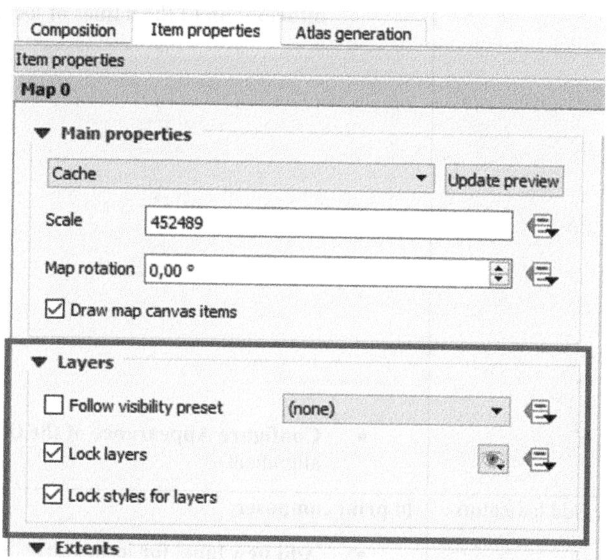
6. Finalize the main map	In print composer: • Add map items: Arrow, scalebar, source, legend

Table 1.12. *Steps for the activation of atlas in print composer*

1.3.4.2. *Add dynamic items*

At this stage, it is necessary to configure the dynamic elements that will be displayed in the atlas maps according to the local regions (project name and indicators).

<table>
<tr><th>Process</th><th>Handling in QGIS</th></tr>
<tr><td>1. Add Title</td><td>
In print composer:
<ul><li>Add new label for the title of template (name of local region) </li></ul>
In Item properties:
<ul><li>Click on Insert an expression</li></ul>
In window:
<ul><li>Indicate the name of the field where we will find the information of the names of local regions</li></ul>
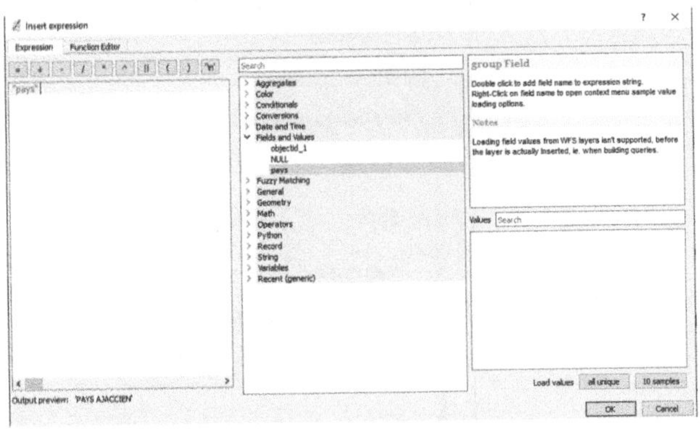

<ul><li>Configure Appearance of the title (Font, font color, size, alignment)</li></ul>
</td></tr>
<tr><td>2. Add indicators</td><td>
In print composer:
<ul><li>Add new label for indicators </li></ul>
In Item properties:
<ul><li>Click on Insert an expression</li></ul>
In window:
<ul><li>Indicate the name of the field where we will find the information of the names of local regions</li></ul>
In the main properties of the indicator label:
<ul><li>Indicate, after the expression, the unit of the indicator</li></ul>
</td></tr>
</table>

	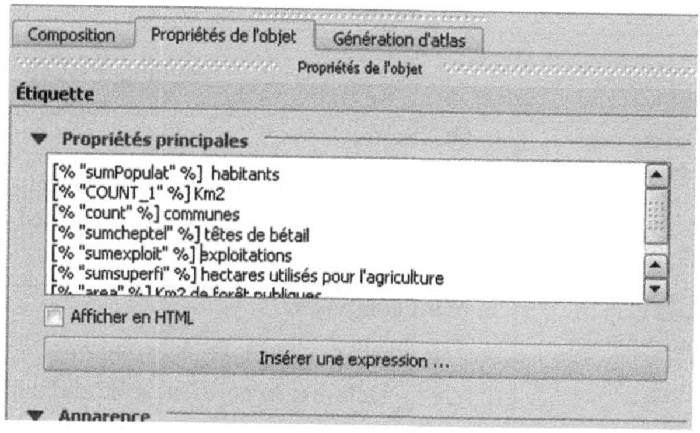 Change the appearance (size, font,…)
3. Add icons	➔ **To give an infographic style to the indicators you can add icons after each indicator (http://www.flaticon.com):** In **print composer**: • Add image 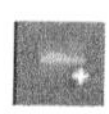 • Define the location of the image (use the SVG format for better rendering) 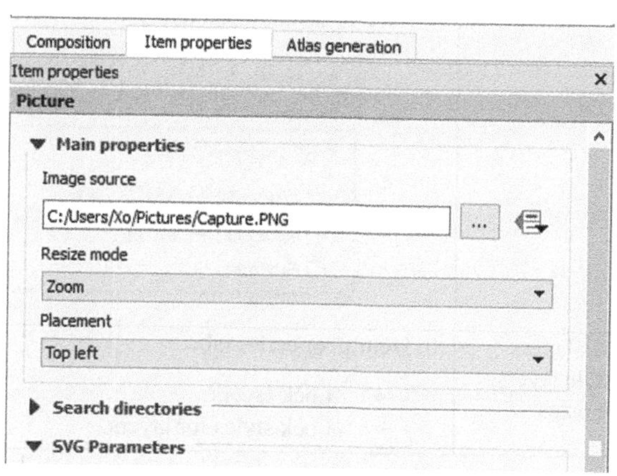

Table 1.13. *Steps for adding dynamic items*

1.3.4.3. *Add municipal population density map*

Process	Handling in QGIS
1. Define map symbology	In **QGIS:** • Activate *municipalities V2* layer • Use a good symbology (graduated style) for representing these data
2. Add map in print composer	In **print composer:** • Click Add Map • Define map position, scale and dimensions
3. Configuration of map	In **Item properties** tab: • Activate the function Controlled by atlas • Configure Margin around feature 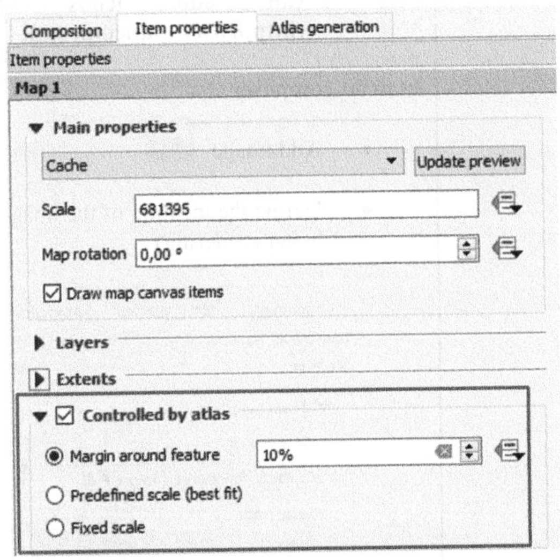
4. Lock layers and style layers of the maps	In **Item properties** tab: • Lock layers • Lock styles for layers

	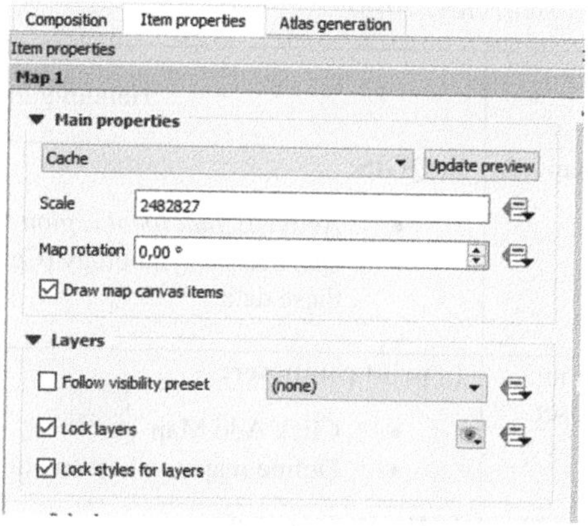

Table 1.14. *Steps to configure municipal population density map*

1.3.4.4. *Add Intermunicipal cooperations map*

Process	Handling in QGIS
1. Define map symbology	In **QGIS**: • Activate *Intermunicipal cooperation* layer • Use a good symbology (Categorized style) for representing these data
2. Add map in print composer	In **print composer**: • Click Add Map  • Define map position, dimensions and scale
3. Configuration of map	In **Item properties** tab: • Activate the function Controlled by atlas • Configure Margin around feature
4. Lock layers and style layers of the maps	In **Item properties** tab: • Lock layers • Lock styles for layers

Table 1.15. *Steps to configure intermunicipal cooperations map*

1.3.4.5. *Add overview map*

<table>
<tr><th>Process</th><th>Handling in QGIS</th></tr>
<tr><td>1. Define map symbology</td><td>In QGIS:
• Activate final local region layer
• Use a basic symbology (Categorized style) for representing these data</td></tr>
<tr><td>2. Add map in print composer</td><td>In print composer:
• Click Add Map 
• Define map position and dimensions</td></tr>
<tr><td>3. Configure overview parameters of the map</td><td>In Item properties:
• Add an Overview 
• Configure this overview with a map frame of Map 0 (the main map)
• Lock layers
• Lock styles for layers
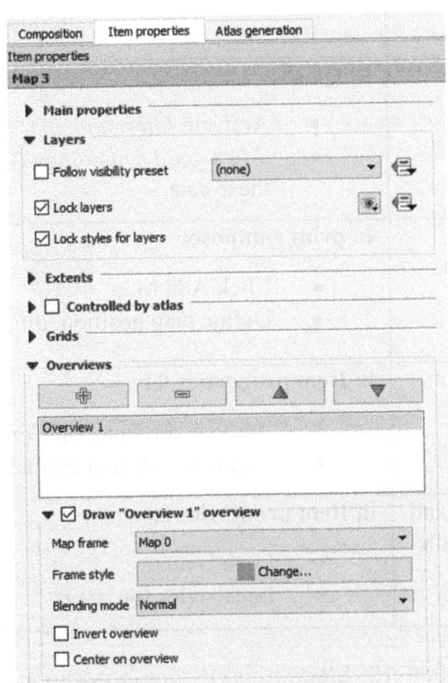
</td></tr>
</table>

Table 1.16. *Steps to configure the overview map*

1.3.5. *Step 5: atlas publication*

The different elements of the atlas are configured (maps, indicators, title, etc.). In this last step, the production of the atlas boards is carried out by exporting them as images.

<table>
<tr><th>Process</th><th>Handling in QGIS</th></tr>
<tr><td>1. Customize output filename with expression</td><td>

In the **Atlas generation** tab:

- Configure the output filename by assigning the value of the field “pays” (name of local regions)

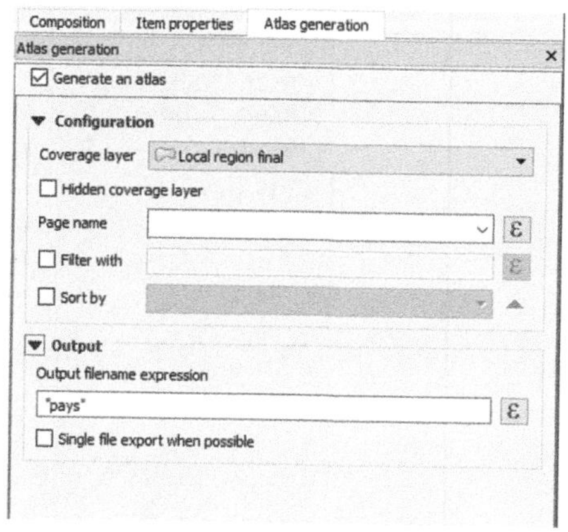

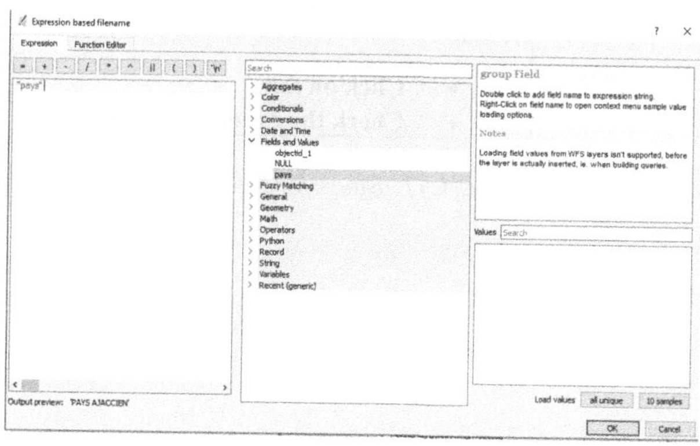

NB : If the field “pays” as the name of files generated by the atlas causes an error under GNU / Linux, then either change the values that caused the problem in the layer or change the output file name by putting the expression as the following: ("pays", "/", "-")

</td></tr>
</table>

2. Run export	In **print composer**: • Click on **Export Atlas as Images** 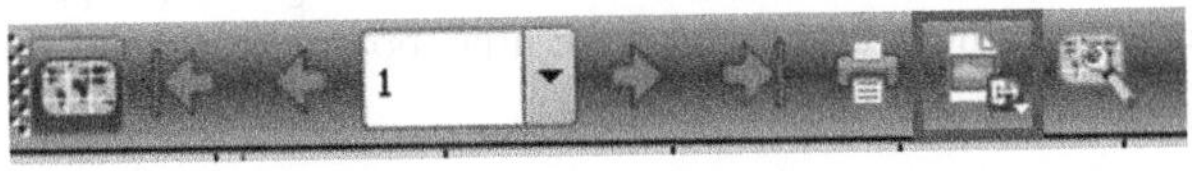 • **Create a folder** in which the pages (images) of the atlas will be stored • Select this directory and choose
3. Configure image export options	**Choose the export resolution** according to the end use of the atlas 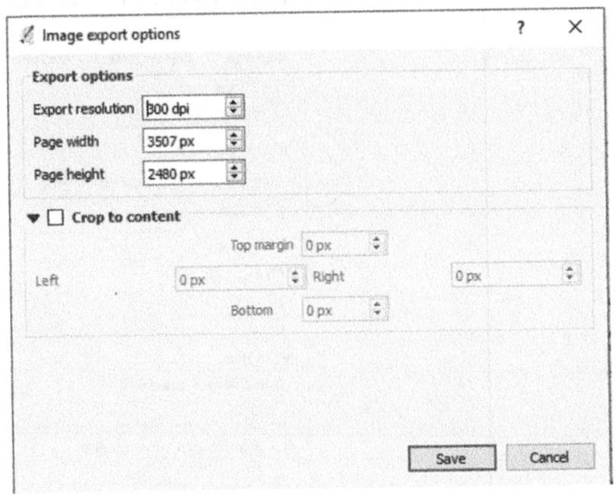 Web publication 100dpi; PDF Publication 300dpi; Printing minimum 400dpi. • **Click on Save** • **Check the export**

Table 1.17. *Steps for the publication of the atlas*

2

Estimation of Land Use Efficiency from the Global Human Settlement Layer (GHSL)

2.1. Context

With the UN's adoption of the 2030 Agenda for Sustainable Development (September 25, 2015, New York)[1], the member states are invited to evaluate their performance with regard to the 169 targets set within the 17 Sustainable Development Goals (SDGs).

The key to the success of achieving these goals lies in the provision of consistent, reliable and regularly updated data to measure the progress in the implementation of SDGs.

Currently, the Global Human Settlement Layer (GHSL), produced by the Joint Research Center (JRC) of the European Commission, represents a rich source of detailed and public access data on human settlements and population on a global scale [PES 16]. The characteristics and description of the GHSL data are presented in the following box.

Chapter written by Christina CORBANE, Panagiotis POLITIS, Martino PESARESI, Thomas KEMPER and Alice SIRAGUSA.

1 http://www.un.org/sustainable development.

Global Human Settlement Layer

The Global Human Settlement Layer (GHSL) is a set of tools and methods for assessing the presence of humans on a global scale.

In the context of GHSL, a global layer on built-up surfaces (GHS-BU) was produced from Landsat image collections for four different periods (1975/1990/2000/2015).

The information generated with the GHS-BU was used to downscale population and estimate population (GHS-POP) for the 1975, 1900, 2000 and 2015 periods.

Data on built-up areas and population can be downloaded for free from the GHSL website: http://ghsl.jrc.ec.europa.eu/data.php.

Box 2.1. *Characteristics of the Global Human Settlement Layer*

2.2. The Land Use Efficiency (LUE)

The GHSL provides the baseline data required to meet the challenges outlined in the SDGs. It does so thanks to the possibility of estimating a series of indicators linked to the evolution of built-up surfaces and population.

In particular, the GHSL offers the potential of specifically monitoring the SDG 11.3[2] target that by 2030 aims to "enhance inclusive and sustainable urbanization and capacity for participatory, integrated and sustainable human settlement planning and management in all countries".

To measure this target, the indicator 11.3.1 – *the ratio between the land use growth rate and population growth rate* – is proposed.

This indicator allows for the measuring of the LUE, which consists of the analysis of the evolution of the relation between the use of land and the growth of population.

However, this indicator is not fully able to grasp the dynamic nature of the cities that show negative or zero negative population growth, nor of those that have lost part of their territories due to a natural catastrophe or conflicts. Moreover, an "urban area" is currently broadly defined as the surface that includes built-up areas and open urban spaces. A consensus on what defines an urban area and its delimitation is necessary in order to refine and apply the aforementioned indicator, due to the variety of the existing definitions and methods.

2 http://www.un.org/sustainabledevelopment/cities.

To overcome these difficulties, we propose to adapt the formulation of the LUE indicator in order to measure the *change rate of the built-up area per capita* [COR 16]:

$$Id_{t,t+n} = \frac{Y_t - Y_{t+n}}{Y_t} \quad [2.1]$$

The built-up area per capita Y_t is defined by:

$$Y_t = \frac{BU_t}{POP_t}$$

where BU_t and POP_t are, respectively, the built-up surface and the population at the time *t*.

The LUE indicator can be estimated at different time intervals (i.e. every 5, 10, 15 years) depending on the availability of the observations. In order to ensure the comparability of the results over time, it is recommended to normalize the values to obtain the *10-year average change*, the equivalent of dividing the indicator by *n (the number of years that separate the observations)* and then multiplying it by 10. The formula of the normalized indicator is:

$$Idx_t = \frac{Y_t - Y_{t+n}}{Y_t} \cdot \frac{10}{n} \quad [2.2]$$

The availability of information on built-up surfaces and populations with different dates offers an interesting case to test the proposed indicator and identify the reasons for changes at different scales from the agglomeration to the regional scale. For this analysis we propose to use the GHSL layer (GHS-BU) and the associated population data (GHS-POP) at a resolution of 250 m × 250 m for the years 1990, 2000 and 2014.

2.3. Installation of the LUE indicator calculation tool

A prerequisite for the application illustrated in this chapter is the installation of the algorithm for calculating the LUE indicator, developed by the JRC team http://ghsl.jrc.ec.europa.eu/tools.php. This compiled script has to be installed in the toolbox only once.

This is compatible with QGIS v2.14 and any posterior versions.

We will use the *Toolbox*, which is the main element of the QGIS processing module and the one that the user will use most frequently. In the toolbox, we can find algorithms classified by category. These categories are classified under the

"Geoalgorithms" group. There are also two other entries: *Templates* and *Scripts*. They contain user algorithms and define processing strings:

– To load the LUE process, simply click on "*Processing*" in the main menu and then select "*Processing toolbox*".

– From the dialogue window, go to "*Script*" → "*Tools*" → "*Add a script from file*".

– Load the python script supplied with the database of this application: *LUE.py*

– Once loaded, the LUE algorithm will appear in the "User Scripts" submenu.

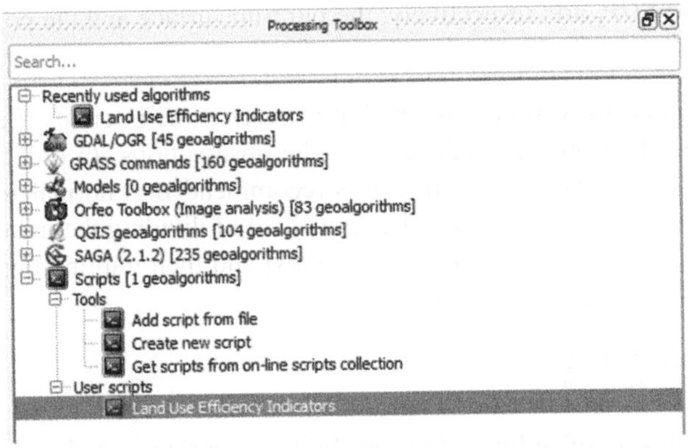

Figure 2.1. *Display of the LUE indicator calculation tool in the QGIS Toolbox*

In the following sections, we will illustrate the step-by-step method of calculating the indicator and its direct implementation in the *LUE.py* tool.

2.4. Method to calculate the LUE indicator

The method of calculating the LUE indicator is based on a geospatial and temporal analysis of raster data. The calculation does not require any preliminary knowledge in remote sensing, however, basic knowledge in the handling and processing of geospatial data is required. The developed method is adapted to the free software QGIS so that any GIS analyst, without special expertise in remote sensing, can use it. Figure 2.2 summarizes the succession of steps required for the calculation of the LUE indicator. To illustrate this method, an example of application on the city of Pretoria in South Africa is presented:

1) Downloading of GHSL data: (i) raster data on built-up areas and (ii) population raster data at a resolution of 250 m × 250 m for the years 1990, 2000 and 2014.

2) Delimitation of the area of interest and clip of input data.

3) Calculation of the built-up area per capita and estimation of the LUE indicator for two different time periods ($Id_{1990\text{–}2000}$ and $Id_{2000\text{–}2014}$).

4) Visualization of the results including a classification of the LUE values and an attempt at interpretation and comparison.

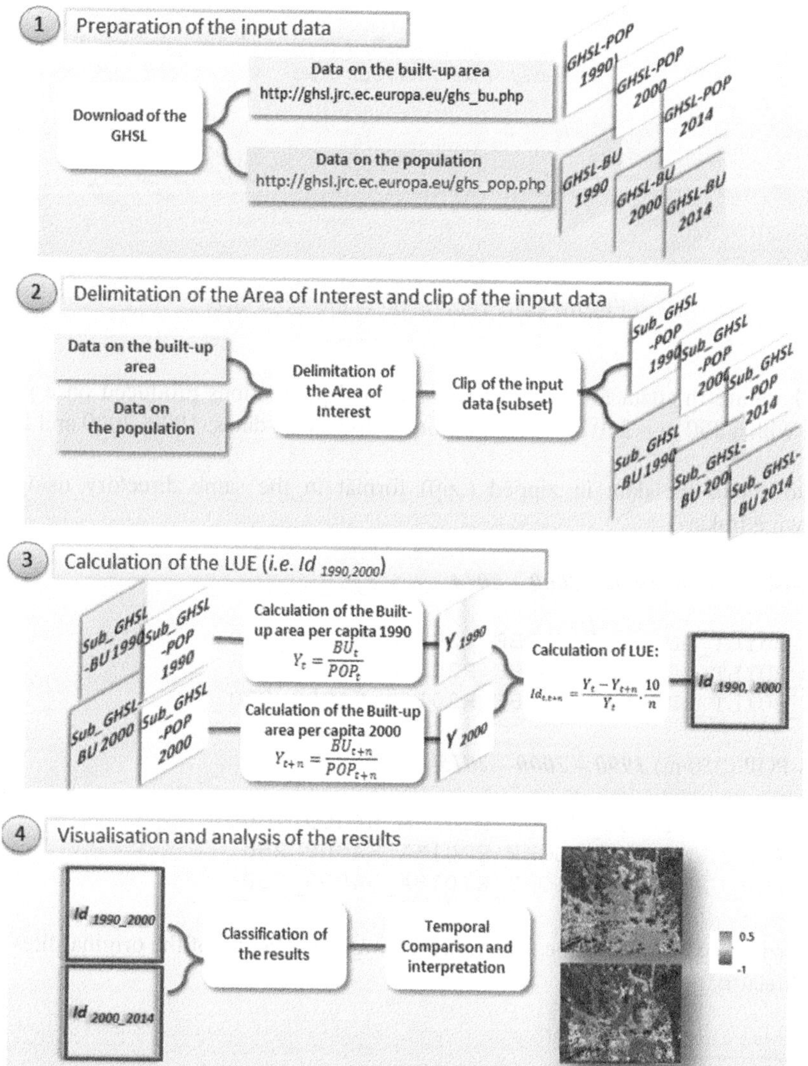

Figure 2.2. *Steps for the calculation of the LUE*

2.4.1. *Preparation of the input layer*

The first step is to download the raster data from the GHSL Website: http:// ghsl.jrc.ec.europa.eu/data.php.

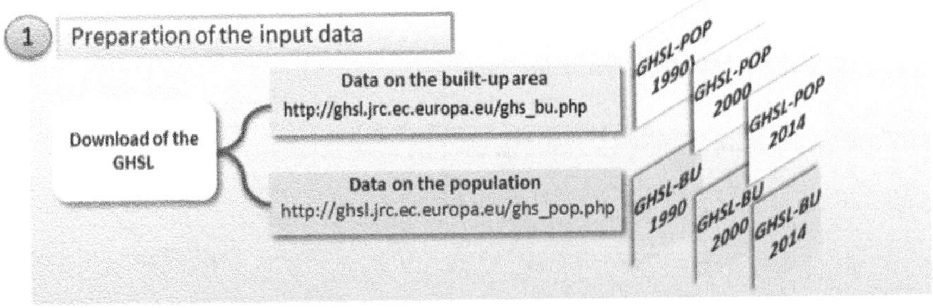

Figure 2.3. *Preparation of the input layer*

These are the data for the built-up areas and populations produced on a global scale with a 250 m × 250 m resolution for the following dates: 1990, 2000 and 2014.

Download the data in zipped (.zip) format in the same directory using the following links:

GHS-BU (250 m) ***1990 – 2000 – 2014***

```
GHS_BUILT_LDS1990_GLOBE_R2016A_54009_250
GHS_BUILT_LDS2000_GLOBE_R2016A_54009_250
GHS_BUILT_LDS2014_GLOBE_R2016A_54009_250
```

GHS-POP (250 m) ***1990 – 2000 – 2015***[3]

```
GHS_POP_GPW41990_GLOBE_R2015A_54009_250
GHS_POP_GPW42000_GLOBE_R2015A_54009_250
GHS_POP_GPW42015_GLOBE_R2015A_54009_250
```

Unzip the file in the same folder, preserving the structure of the original files (i.e. one directory per data).

3. Since the population layer for 2015 is produced from the built-up layer extracted from the Landsat 2014–2015 collection, for simplification and consistency reasons, we named the layer: GHSP-POP 2014.

2.4.2. *Delimitation of the area of interest and clip of input data*

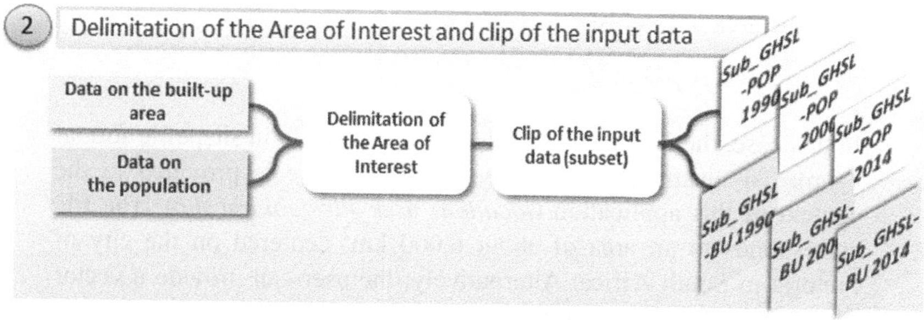

Figure 2.4. *Delimitation of the area of interest and clip of input data*

Once the input data are available, it is possible to launch the LUE process by clicking on the tool previously loaded in the user scripts.

The following window (Figure 2.5) appears.

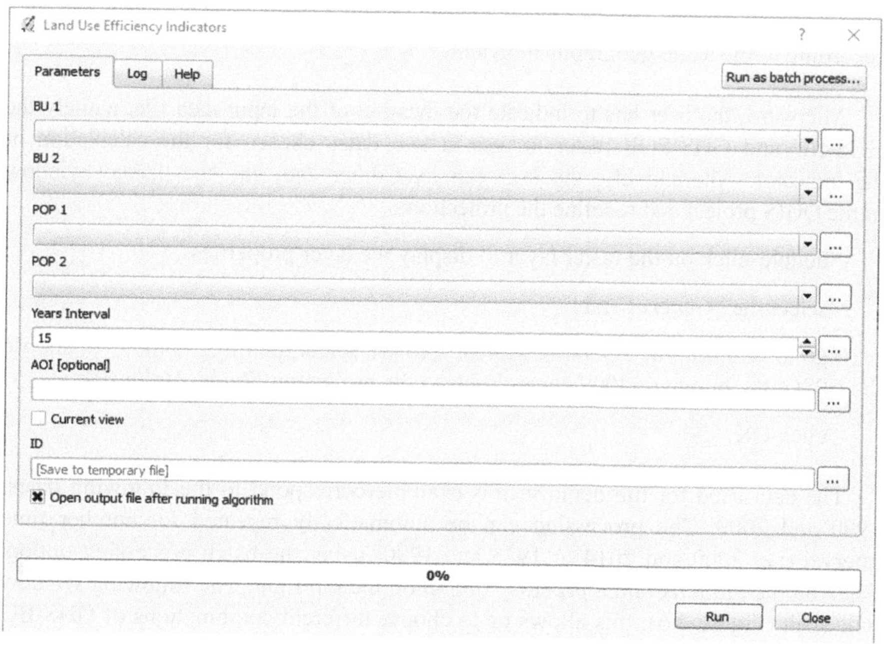

Figure 2.5. *The interface of the LUE tool*

Step 2 of the processing chain consists of defining an area of interest and cutting the input data onto the previously delimited area of interest. There are two options for defining the area of interest:

– *Option 1: AOI [Optional].*

In this case, the area of interest is defined by a file in shapefile format (.shp). To illustrate this approach, a vector file is provided in the context of this application (*location: user_files/aoi/aoi.shp*). The file corresponds to an area of about 6,000 km^2 centered on the city of Pretoria in South Africa. Alternatively, the user can provide a vector file delimiting a chosen study area.

– *Option 2: Current extent.*

Another possibility is to use the current extent of the window as the area of interest. To do this, simply check the box *corresponding* to Current Extent.

Whichever option is chosen, the definition of the area of interest will be used to extract a subset of the data to be analyzed: all input raster images will be divided according to the same geographical extent.

Afterward, the user has to indicate the location of the input data file, namely the GHS-BU and GHS-POP raster images at two dates chosen for the calculation of the indicator. Alternatively, the user can load other data into the display window of the QGIS project and redefine the projection:

– double-click on the raster layer to display the layer properties;

– select the "*General*" tab;

– go to "*Coordinate reference system*" (CRS), select the filter "54009" to choose the EPSG: the number 54009 corresponds to the projection *World_Mollweide*;

– click OK.

The data used for the demo in this example correspond to the following dates: 1990 and 2000. The processing can be automatically repeated for another time interval (i.e. 2000 and 2014 or 1975 and 1990) using the batch processing option. Click on the "*Run as batch process*" option on the top right. The following window will open (Figure 2.6); this allows us to choose different combinations of GHS-BU and GHS-POP for the desired LUE indicator, as well as per different areas of interest.

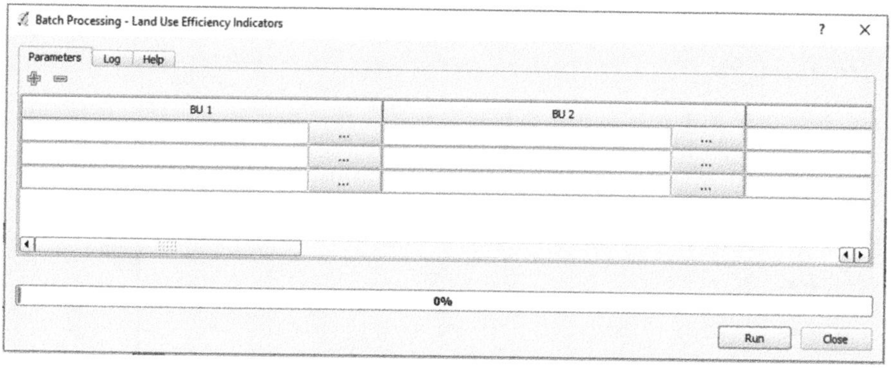

Figure 2.6. *Batch process execution window*

2.4.3. *Calculation of the LUE indicator*

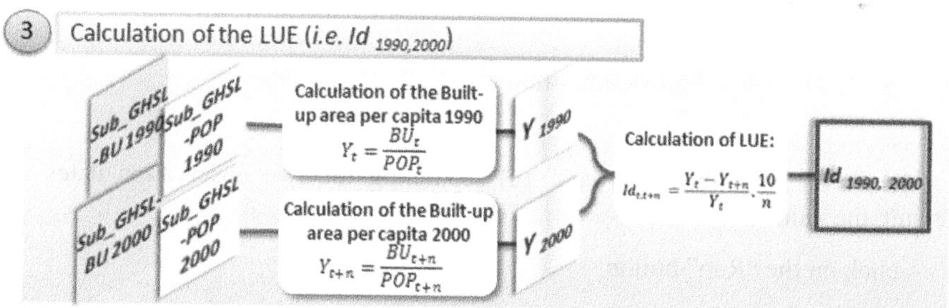

Figure 2.7. *Calculation of the LUE*

– The user can choose to save the output result either in a temporary file or locally, by choosing the location and format of the output file:

- under the "*LUE*" option of the *Land Use Efficiency* user interface, click on the three points [...];

- choose one of the two backup options that appear;

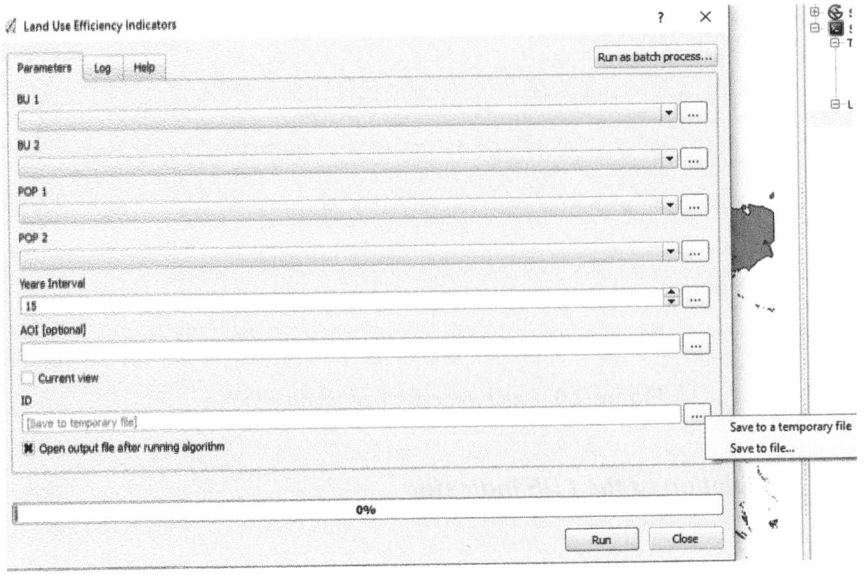

Figure 2.8. *Choosing the backup option*

– enter the "Number of Years" (n) separating the two observation dates. By default, the value is $n = 15$;

– click on the "Run" button;

– the algorithm will then calculate, for each of the dates, the built-up surface per capita $Y_t = \frac{BU_t}{POP_t}$ and $Y_{t+n} = \frac{BU_{t+n}}{POP_{t+n}}$. Thereafter, the normalized indicator Idx_t is estimated according to equation [2.2];

– the side menu will be displayed in the window if the option "Open output file after running algorithm" is checked;

– we repeat steps 1, 2 and 3 for dates 2000 and 2014 to calculate indicator $Id_{2000\text{–}2014}$. Alternatively, it is possible to simultaneously calculate the indicator for both time intervals with the option "Run as batch process".

2.4.4. *Visualization and analysis of the results*

Once the results of the indicator are ready and displayed in the QGIS project, it is time to visualize them and try to interpret the values of the SGD.

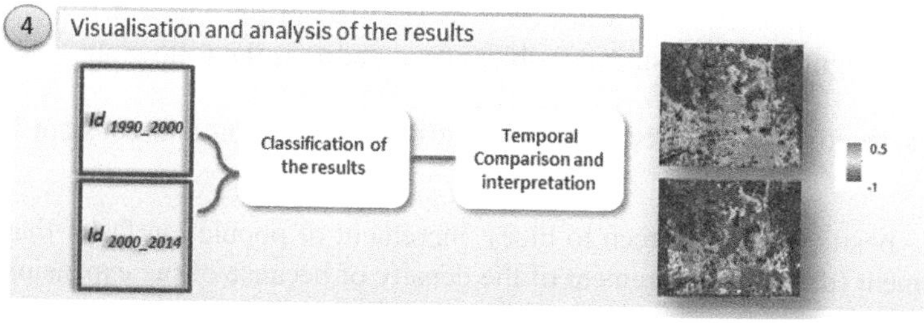

Figure 2.9. *Visualization and analysis of the results*

In order to facilitate the interpretation of the values, one style of visualization and classification is proposed; we suggest changing the style in the following way:

– double-click on the name of the raster file corresponding to the indicator (i.e. $Id_{2000-2014}$) in the key or right-click on the name and choose "Properties" from the menu that appears. The Layer property window will appear;

– go the Style tab and click on the "Style" button on the bottom right to change the style proposed in this example in order to facilitate the interpretation of the values. The proposed visualization style *LUE_style.qml* is located in the folder provided for this application (*location: user_files/LUE_style.qml*)

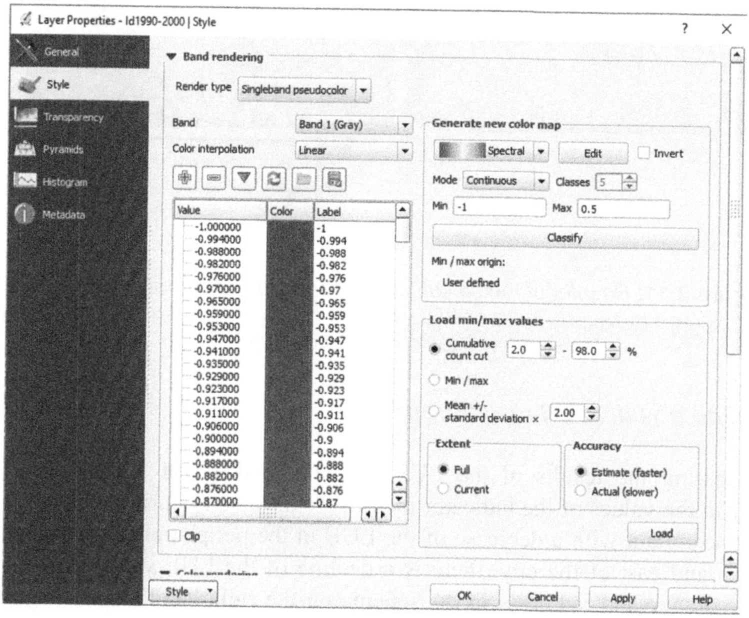

Figure 2.10. *Layer property window in QGIS*

– In this color palette, the values are reclassified in the interval [–1, +0.5], in order to highlight the contrast. A possible interpretation is the following:

- negative values (red to orange): loss of population and constant built-up surface;

- positive values (green to blue): increment of population faster than built-up increment (due to the increment of the density or because of the expansion of the urban area);

- values around zero (yellow): stable zones with a linear increment of built-up surfaces and population.

With the example of the city of Pretoria, Figure 2.11 shows the results of the calculation of the LUE indicator for two intervals: $Id_{1990–2000}$ and $Id_{2000–2014}$.

Figure 2.11. *Results of the calculation of the LUE indicator for two intervals: $Id_{1990–2000}$ and $Id_{2000–2014}$ (cells of 250 m × 250 m). For a color version of the figure, see www.iste.co.uk/baghdadi/qgis3.zip*

2.4.5. *One possible interpretation*

Comparing the results of the LUE of the two different periods, there is an increase in the values of the indicator in the core of the city during the period 2000–2014 concomitant with a decrease of the LUE in the peripheral areas. Especially in the north and east of the city, there is a decline of the LUE values corresponding with the orange and red areas in the schema on the right-hand side of Figure 2.11. This could be due to the migration of a population from the peripheries toward the center.

On the other hand, there is a general increase in the values of the LUE indicator between the period 1990–2000 and 2000–2014, particularly in the area represented in Figure 2.2 by an ellipse. A deeper analysis of this area shows that it corresponds to a typical area of the urban landscape of the city of Pretoria where social housing was built by the government in the last ten years in response to the housing crisis. However, this solution did not solve the housing problem for the poor population. In fact, this landscape is now characterized by over-occupancy of existing formal dwellings and by the construction of backyard shacks, an example of which is given in Figure 2.12 for the area represented by the ellipse.

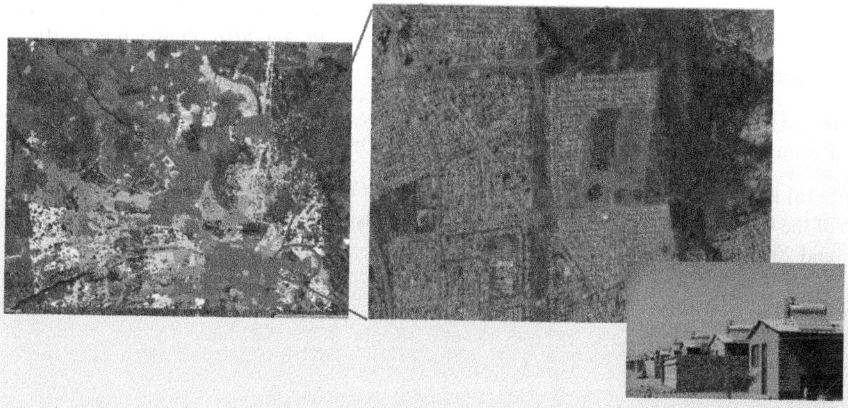

Figure 2.12. *Example of backyard shacks that are formal government-built housing, but where over-occupation corresponds to an average of 15 people living in a four-room dwelling [GER 13]. This explains the increase in the value of the LUE indicator in these areas. For a color version of the figure, see www.iste.co.uk/baghdadi/qgis3.zip*

2.5. Limits of the method

The method for calculating the LUE indicator from GHSL data can be easily applied to any datasets on built-up areas and population (i.e. more detailed local data or data at the municipal or regional level). It can be adapted to the monitoring purposes of the LUE indicator. It also provides a source of comparative information in time and space (e.g. between cities) that is essential for assessing the impact of local urban planning policies.

Its main limitation lies in the difficulty of grasping the vertical development of constructions due to two factors: (1) at the moment, the input data represent 2D information of built-up surfaces and populations, and (2) the proposed indicator does not take into account information concerning the volume of cities, because of the difficulties in obtaining this information at a global scale.

Among the developments envisaged within the framework of GHSL, there is the provision of information on the volume of built-up areas on a global scale. This will be used for a more precise modeling of population data and to improve the estimation of the LUE indicator.

2.6. Bibliography

[COR 16] CORBANE C., PESARESI M., KEMPER T. *et al.*, "Assessment of Land Use Efficiency using GHSL derived indicators", *Atlas of the Human Planet 2016*, Publications Office of the European Union, pp. 82–83, 2016.

[GER 13] GERVAIS-LAMBONY P., "Chapitre 4-Le Fait Urbain En Afrique Du Sud: Recomposer Les Structures Héritées, Donner Le Droit À La Ville", pp. 73–90, 2013.

[PES 16] PESARESI M., EHRLICH D., FERRI S. *et al.*, "Operating procedures for the production of the Global Human Settlement Layer from Landsat data of the epochs 1975, 1990, 2000, and 2014", Joint Research Centre, 2016.

3

Characterizing Urban Morphology for Urban Climate Simulation Based on a GIS Approach

3.1. The city–climate relationship through urban climate modeling

Urban areas are original and complex as they are characterized by diversified components that interact with one another. In particular, many studies show the impacts of the urban form and design on the local climate [OKE 73; ESC 91; MAS 15]. The urban climate simulations are specific as some characteristics have to be taken into account at a fine spatial scale, like, for instance, the building's size, form and configuration, which imply changes in the atmospheric flow regime (from laminar to turbulent) as well as shadow effects. Furthermore, the different materials used in urban constructions (bitumen, cement, tiles, slates, aluminum sheets, etc.), which have various physical properties, modulate the albedo, and the surface conductivity that, in turn, change the surface radiation balance. The resulting urban climate, for which construction materials, urban morphology (at local scales) and regional topography play a key role, is thus different from the surrounding natural environment climate.

The present work was achieved in the framework of the MUSTARDijon interdisciplinary[1] research program, dedicated to the Urban Heat Island (UHI) description in the city of Dijon. This research program is based on three approaches

---

Chapter written by Justin EMERY, Julita DUDEK, Ludovic GRANJON, Benjamin POHL, Yves RICHARD, Thomas THEVENIN and Nadège MARTINY.

1 The MUSTARDijon program (*Mesures Urbaines de la Température de l'Air à Dijon* : *Urban* Climate *measurement* in Dijon, Bourgogne, France) brings together researchers from GIS, climatology and remote sensing.

to determine the temperature variable: climate modeling, *in-situ* measurements and remote sensing products. The Geographical Information System (GIS) plays a central role in this study and is used to process the land use database (urban morphology, land use, buildings height) in order to ensure the temperature measurement network implantation or process the remote sensing products. The GIS is also used to construct a high resolution database that feeds a Mesoscale Atmospheric Model (MAM). In the latter case, the land use database is necessary to solve the surface–atmosphere exchanges. In the MAMs, the meteorological conditions come from the atmospheric forcing from a General Circulation Model (GCM). Indeed, the modeling of the city–climate relationship requires quantification of: (1) the thermic exchanges due to anthropization and buildings' morphology (height and position); (2) their impacts on heat fluxes and the associated surface radiation balance [RYU 11]. Thus, urban climate modeling requires characterization of the urban morphology and the study area topography. In our case, there are two key parameters: the distribution of anthropized surfaces and that of the buildings within the urban area. Geographical databases available for many cities across France enable us to get an overall picture and a realistic description of urban morphology. In our case, four databases were used: BD TOPO®[2] and BD ALTI®[3], produced by the French National Institute of Geographic and Forestry Information (IGN), completed with the Corine Land Cover[4] (CLC) database for the land use and OpenStreetMap[5] (OSM).

The aim of this chapter is to formalize a geographical processing chain using general databases and free GIS software (here Quantum GIS) in order to construct MAM input datasets and simulate an urban climate. In this study, the spatial resolution ranges between 150 m and 1 km, in agreement with the MAM's resolution defined as a good compromise in terms of calculation time and urban climate observation at a relatively fine spatial scale. Given these elements, two steps must be achieved:

1) **Integrate the buildings' heights into the Digital Terrain Model (DTM)** in order to achieve the best results in modeling the thermic and atmospheric processes related to the building's presence.

---

2 The BD TOPO® database provides a vector description of the territory allowing for description and vizualisation of the geographic objects in 3D (transport infrastructures, buildings…) [IGN 16].

3 BD ALTI® database is a Digital Terrain Model (DTM) to describe the terrain at a medium scale (from 25 to 250 m).

4 European cartography of the land use by Corine Land Cover: http://land.copernicus.eu/pan-european/corine-land-cover/clc-2012/view.

5 In the practical case in section 3.3, only open-databases are used, excluding the manipulation of the BD TOPO® database.

2) **Provide an overall picture of "anthropization"**, which makes it possible to distinguish the "natural" areas from the "artificial" areas and characterize their influences on thermic conditions.

In France, the anthropization concept directly refers to a "*fact, provoked, accentuated by human action*" [BRU 05]. In our study region, the modifications of natural areas most frequently result from anthropization, such that "*the environment is more the work of the human than of nature*" [VEY 97]. The term "anthropization" corresponds to the urban fabric, constituted of urbanized areas referred to as "artificial" areas and comprising built-up and none built-up surfaces. Two definitions of the urban fabric have to be distinguished [CER 13]:

– **The continuous urban fabric** (for interurban problems): corresponds to surfaces structured around buildings and communication routes covering more than 80% of the total area of a city.

– **The discontinuous urban fabric** (for the intraurban problems): corresponds to surfaces structured with buildings, highway and artificial areas with vegetal surfaces and bare soils covering significant areas of the city. In this case, the urban elements cover between 30% and 80% of the total area of a city.

Purpose	Characterizing urban morphology for the numerical urban climate simulation
Spatial resolution	Medium: from 150 m to 1 km
Database product	Digital terrain model, land use, transport infrastructures, building's heights
Coordinate reference system	Lambert 93 – French Geodesic Network 1993 *CRS: RGF93/Lambert-93; ID Certified EPSG*[6]*: 2154*
Expected results	Urban topographic context; anthropization index at the urban scale

Table 3.1. *Synthesis of criteria for the characterization of urban space for the numerical simulation of the urban climate*

The method proposed here for the estimation of anthropization corresponds to the calculation of a ratio between two surfaces: the surface of the urban area by considering a discontinuous urban fabric assumption and the total available surface. This ratio is referred to as the anthropization index in the following. Note that in the following the terms "anthropization" and "artificialization" are both used to denote areas with important human constructions (buildings, transport infrastructure, etc.).

6 EPSG (*European Petroleum Survey Group* is the global structured dataset of Coordinate Reference Systems and Coordinate Transformations: http://www.epsg.org).

In summary, we propose obtaining an image of anthropization through the definition of the anthropization index based on GIS software and geographical databases. The geometrical information provided by the geographical database makes it possible to acquire good estimations of the buildings' footprints and heights, transport infrastructures and natural areas such as urban park footprints.

3.2. Representation of the urban space

In the following sections, we detail the processing chain that makes it possible to provide a 3D image of the city. For convenience, the different phases of the processing are gathered into three steps (Figure 3.1):

– **Step 1**: Integrate the buildings' heights into a DTM.

– **Step 2**: Define the anthropization within the study area.

– **Step 3**: Calculate the anthropization index.

Prior to achieving these three steps, we define the reference coordinates system. Here we used the Lambert 93-RGF93 (CRS: RGF93/Lambert-93; ID Certified EPSG: 2154) projection, which is the most commonly used metric coordinates system in France.

QGIS functionality to reproject a raster layer:
- Changing native projection: *Raster > Projections > Wrap (Reproject)...*

QGIS functionality*[7] to reproject a vector layer:
- Changing native projection: *QGIS geoalgorithms > Vector general tools > Reproject layer*

The processing chain uses reference vectorial databases from the IGN databases and/or OSM databases. The aim is to develop a simple and generalistic approach which could be used in other cities. QGIS software with basic functionality (without extensions) is used to build up the whole processing chain. Note that this work could be done with another GIS software like ArcGis or MapInfo. However, it is noteworthy that this processing chain results from feedbacks obtained in the framework of the MUSTARDijon program and based on the skills of climatologists and engineers in geomatics. Our approach is based on vectorial databases.

7 The QGIS functionalities in the processing Toolbox are asterisked: *. Click on Processing > Toolbox ; or use the following keyboard shortcut: *Ctrl+Alt+T.*

Supplementary raster files (remote sensing datasets) could also be used, as illustrated at the end of the chapter. Three complementary vectorial databases are used in this work:

– **A TOPOGRAPHY description of the territory provided by the BD ALTI database** that is an open-data DTM at a 75 m spatial resolution: http://professionnels.ign.fr/bdalti#tab-3.

– **A LAND USE description** is provided by the CLC database at a 100 m spatial resolution that is distributed by the European Environment Agency: http://www.eea.europa.eu/data-and-maps/data/corine-land-cover-2006-raster-2.0

– **A 3D TERRITORY and TRANSPORT INFRASTRUCTURES description** is provided by the BD TOPO® database. This is the source of a large amount of information (*administrative structure, points of interest, toponymy, road, railway, energy, hydrologic networks, buildings, vegetation, and orography*). For our study, we selected the following thematic files:

- *roads network: roads and road surfaces;*

- *railways network: railways sections;*

- *buildings and their heights: undifferentiated buildings (Industrial or Outstanding);*

- *vegetation: vegetation surface;*

- *hydrographic: hydrographic surfaces;*

- *optionally, complementary thematic files could be used such as aerodrome runway or activity areas (industrial and commercial).*

Our work is mostly based on the BD TOPO® because of the availability of vectorial 3D territory description. Unfortunately, this database is not an open-source one. Thus, we used, when possible, information from the OSM database even though the information does not have the same reliability and accuracy. As for the CLC databases, the OSM databases give satisfactory information about the land use for Dijon, especially the one related with the identification of the buildings' positions in the tertiary sector areas (see section 3.2.2.2). Furthermore, the BD TOPO® is complementary to OSM for vegetation data processing (see section 3.2.2.3). The steps of the processing chain are detailed in Figure 3.1.

First of all, the building's height, provided by the BD TOPO® based on photogrammetric processes conducted by IGN, is an important parameter to simulate urban climate. Indeed, the roughness of the building's surfaces modifies the natural airflow. Moreover the building's height is useful to identify the "canyon streets", defined as the ratio between the building's height and the street width has to be greater than 0.5. Note that the Sky View Factor (SVF), which quantifies the fraction of sky visible from the ground up, is also a widely used index in urban climatology for canyon streets identification. It can be estimated from the processed information described in section 3.2.1.

The footprint of the urban elements is another key parameter in our processing chain. Indeed, the presence of highly anthropized areas modifies the natural runoff and water penetration as well as the heat flux stored or reflected by various surfaces. As a result, these two parameters are essential and have to be accounted for in the estimation of the anthropization index (third step of the work, see section 3.2).

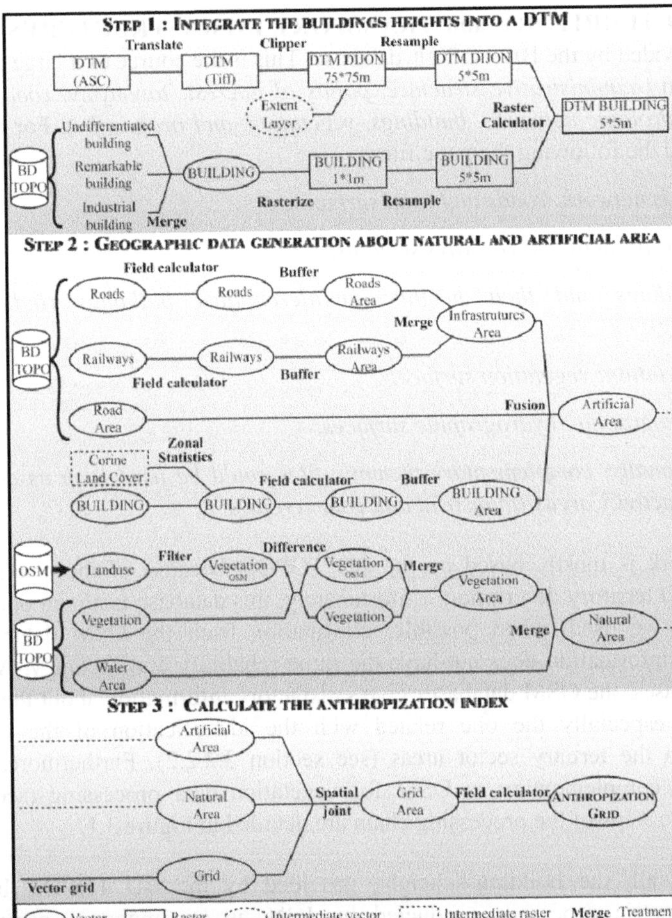

Figure 3.1. *Processing chain used to calculate the anthropization index in Dijon*

3.2.1. *Step 1: integrate urban morphology in a DTM*

This step aims to create an image of the city based on the topographic context (buildings' heights). The presented pre-processings are useful to ensure the interoperability between the BD ALTI® and the buildings' heights extracted from the BD TOPO®.

3.2.1.1. *The BD ALTI® pre-processing: convert and clip the study area*

BD ATI® database is a DTM that describes the shape and elevation of the earth's land surface[8]. Before the DTM processing, its original format (ASC) is converted to GeoTIFF (.tiff). Indeed, even though QGIS is able to open the ASC files, it is easier to work with the classic TIFF raster format:

QGIS functionality to convert a raster (ASC to GeoTIFF):
- Convert data format: *Raster > Conversion > Translate (Convert Format)...*

The next step consists of extracting the study area using the coordinates defined for the city of Dijon (these coordinates are defined in meters to the perimeter contour of the field of study): *xmin = 842701.280, xmax = 867207.099, ymin = 6684746.823, ymax = 6700866.768*. It is also possible to extract the study area using a vectorial layer corresponding to the contour (perimeter) of the domain generally delivered in a shapefile format and known as a "mask layer":

QGIS functionality* (GDAL extend) to extract the study area from vector layer:
- Clip: *GDAL/OGR > Extraction > Clip raster by mask layer*

3.2.1.2. *The BD TOPO® pre-processing: process of building rasterization*

The BD TOPO® offers a 3D description of many objects at a spatial resolution of 1 m. More specifically, the BD TOPO(r) provides buildings' heights obtained based on photogrammetric processes conducted by IGN. Initially, the raw information related to buildings is stored in three separate shapefiles (.shp):

– **INDUSTRIAL BUILDINGS**, which includes all buildings of an industrial, commercial or agricultural type.

– **OUTSTANDING BUILDINGS**, which includes all buildings with administrative, religious, sporting or transport-related functions.

---

8 Download link for the BD ALTI® in open-source (at DTM resolution from 75 to 250 m): http://professionnels.ign.fr/bdalti#tab-3.

– **UNDIFFERENTIATED BUILDING**, which includes all other types of buildings.

The merging of these three shapefiles into one same file enables all the buildings to be gathered in one single layer, previously referred to as "BUILDING":

> **QGIS functionality* to merge vector layers:**
> - Merge: *QGIS geoalgorithms > Vector general tools > Merge vector layers...*

Since the data produced by the IGN is available for an entire region (in the French administrative sense), it is necessary to extract the buildings contained in the study area using the intersection between the previously merged building vector layer and the coordinates of the study area (defined previously):

> **QGIS functionality* to extract vectors from the geographic coordinates of the study area:**
> - Clip: *GDAL/OGR > [OGR] Geoprocessing > Clip vectors by extend...*

The rasterization of the building's height is then based on the geometric information field named "HEIGHT" in the DTM. Figure 3.2 illustrates the conversion of a vector layer to a raster layer at its lowest spatial resolution.

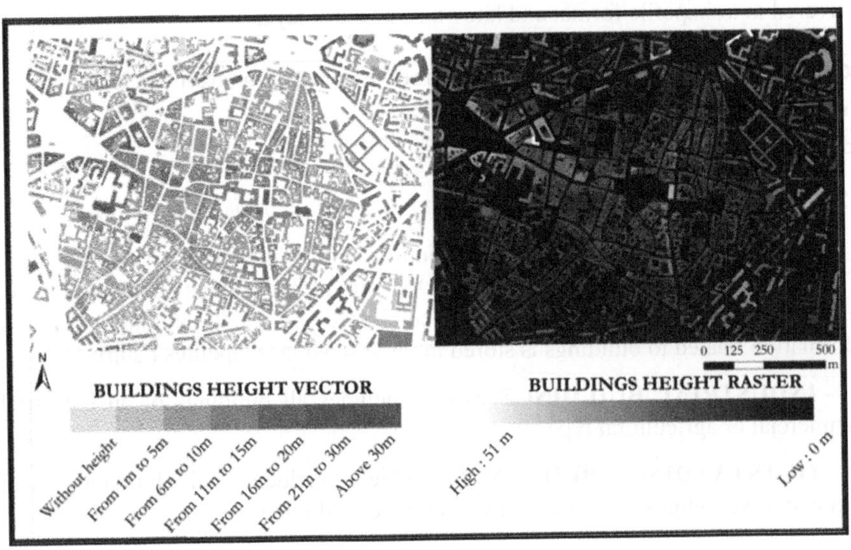

Figure 3.2. *Buildings' heights rasterization process from BD TOPO®. For a color version of the figure, see www.iste.co.uk/baghdadi/qgis3.zip*

QGIS functionality to convert a vector layer to a raster layer:
- Rasterization: *Raster > Conversion > Rasterize (vector to raster)*

3.2.1.3. *Integration of buildings' heights into the DTM: resample and merge raster datasets*

The work consists of combining the buildings' heights with the topography information. In short, two raster files (DTM and BUILDING) are merged into one. Note that it is important to merge raster files at the same spatial resolution. A common resolution of 5 m × 5 m was selected in order to ensure the preservation of the building morphology. Indeed, an oversized aggregate would tend to smooth the buildings' perimeter. Conversely, a too-high resolution would require a long processing time. The original resolution of DTM being of 75 m × 75 m, the spatial resolution of 5 m × 5 m is obtained based on an interpolation method (bilinear). Conversely, the original resolution of the BUILDING file being of 1 m × 1 m, an aggregation process must be applied to the initial raster file (see Figure 3.3).

QGIS functionality* (GRASS extend) to resample a raster:
- Resample using interpolation: *GRASS > Raster > r.resamp.interp*
- Resample using aggregation: *GRASS > Raster > r.resamp.stats*

Once both rasters (DTM and BUILDING) are created at the same spatial resolution, the integration of buildings' heights into the DTM is a simple sum of the elevation values (buildings' heights added to relief topography) pixel-by-pixel using the Raster Calculator tool (one pixel corresponds to an elevation value). The final result (see Figure 3.3) reflects the global topography context of the study area:

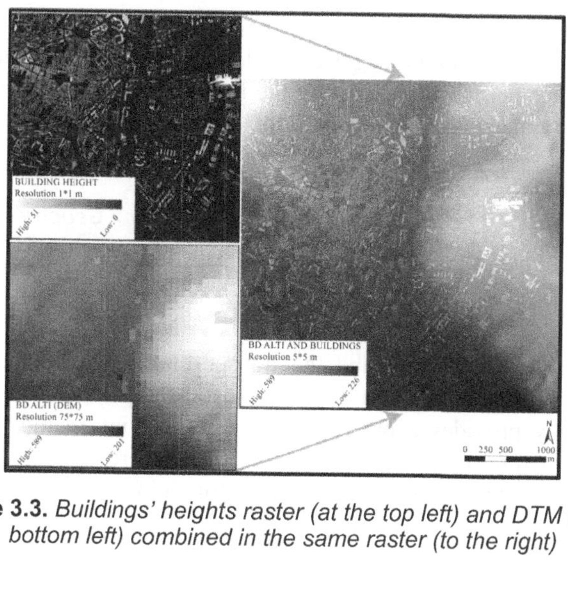

Figure 3.3. *Buildings' heights raster (at the top left) and DTM (at the bottom left) combined in the same raster (to the right)*

QGIS functionality to merge raster layer:
- Merge DTM with buildings height: *Raster > Raster Calculator...*

The new raster file is useful to account for general phenomena, which will be taken into account in the urban climate simulations, such as airflow changes, as well as heat flows and thermal exchanges provoked by the buildings' presence at the street scale [RYU 11]. In addition, this final-DTM makes it possible to lead other complementary analyses such as the study of the shadows effect or the SVF, which are used to settle the temperature measurements network throughout the Dijon territory [ROU 14].

3.2.2. *Step 2: generate land use in an urban space*

The second step consists of exploiting the BD TOPO® to describe the anthropization of the studied urban space (Dijon in this case). BD TOPO® will be used to reconstruct a synthetic map of Dijon by separating artificial (transport infrastructures and buildings) from natural areas. The BD TOPO® vector datasets used are:

– **Transport infrastructures' footprints**: ROADS and RAILWAYS.

– **Buildings' footprints**: INDUSTRIAL_BUILDING, OUTSTANDING_BUILDING, UNDIFFERENTIATED_BUILDING and SURFACE_ROADS.

– **Natural areas' footprints**: VEGETATION_AREA and HYDROGRAPHIC_AREA.

Complementary datasets can be used to refine the surface calculations. In some cases, business parks cannot be neglected and car parks or airport runways have to be considered. Geometric processings on the footprints of transport infrastructures, buildings and natural areas are based on buffer definition using GIS:

QGIS functionality to create varying buffers:
- *Creating buffer from a field in the attribute table: Vector > Geoprocessing Tools > Variable distance buffer*

3.2.2.1. *Pre-processing of the transport infrastructures' footprints*

Here the purpose is to describe the transport infrastructures' footprints that should take the infrastructures' width and realistic footprints into account. As the BD TOPO® only provides a part of the geometrical information related to the transport infrastructures (width and number of lanes), there is a need to complement the available dataset by adding, for instance, information about the sidewalks. This can be done based on local legislation like the town planning code. All the following pre-processings enable us to estimate the railways' and roads' footprints.

3.2.2.1.1. Pre-processing of the railways

The French reference for the gauge of railways is 1.435 m[9] by default. In pre-processing, this value is multiplied by the number of lanes provided in the BD TOPO®. We also consider the railway ballast by defining a distance of 2.5 m around the railways. Thus, the total footprint of each railway is 3.495 m. Note that the footprint of urban railways, such as tramways, is processed in the same way.

QGIS functionality to calculate the width of railways:
- Open attribute table > Open the field calculator (Ctrl-I) ...

3.2.2.1.2. Pre-processing of the roads

The information related to the footprints of roadways in the BD TOPO® database is also incomplete. The width of the roadways provided by the BD TOPO® is not representative of the real footprint of each street (i.e. distance between buildings) but of the distance between lanes. The real width of the streets can be determined either by:

1) **calculating the distance between buildings along each street** without taking the geometric characteristics of the roads (width, number of lanes…) into account;

2) **estimating the width of the street** using the geometric characteristics of the roads to which complementary information is associated.

The choice of the methodology used is crucial. For instance, the shadow effects are not the same based on method 1 or 2. The estimation of the real width of the streets based on method 1 (distance between buildings) tends to overestimate the width of the street and underestimate the shadow effects. This is the contrary effect for method 2. In our case, we apply method 2 for which the sidewalk widths are defined according to the type and number of lanes based on local road legislation[10] (see Table 3.2).

---

9 French official track gauge of railway approved on January 28, 1977 (NF F50-402, February 1977): https://rhcf.revues.org/1939#tocto1n8.

10 Local road legislation from the city of Dijon: https://web2.dijon.fr/shared/images/fichier_joint_00024360_7401.pdf.

Type and number of lane (original BD TOPO French field)	Size of sidewalks
Highway (Autoroute)	0
Link (Bretelle)	0
Path (Chemin)	0
Steps (Escalier)	NC
Cycleway – 1 lane (Piste Cyclable)	2 m
Cycleway – 2 lanes (Piste Cyclable)	2.5 m
Trunk (Quasi-autoroute)	0
Graveled road (Route empierrée)	3.5 m × 2
1-lane road – 1 lane (Route à 1 chaussée)	3.5 m × 2
1-lane road – 2 lanes (Route à 2 chaussées)	5 m × 2
2-lane road – 1 lane (Route à 1 chaussée)	3.5 m × 2
2-lane road – 2 lanes (Route à 2 chaussées)	5 m × 2
Track (Sentier)	0

Table 3.2. *Size of sidewalks according to the number of lanes*

QGIS functionality to calculate the streets width:
- Open attribute table > Open the field calculator (Ctrl-I) ...

3.2.2.1.3. Generation of the transport infrastructures layer

Once the data related to railways and roadways are generated as surface features, we merge the two files in order to obtain one polygonal transport infrastructures layer. However, before this operation, it is important to remove geometric overlaps between layers using the QGIS *difference* processing tool (see Figure 3.4). The generation of the transport infrastructure layer is done following three steps:

1) **Generation of the buffers related to railways and roadways using the infrastructures' width** detailed previously. Note that this step changes the nature of the layer – from polyline to polygonal.

2) **Removal of geometric overlaps between layers**. This is done using the QGIS "*difference*" processing tool and the roadway intersection information (see Figure 3.4).

3) **Merging of railways and roads layers**. This last step enables us to create the geometric features of the transport infrastructures. The final result is a visualization of the footprints of all transport infrastructures in the study area.

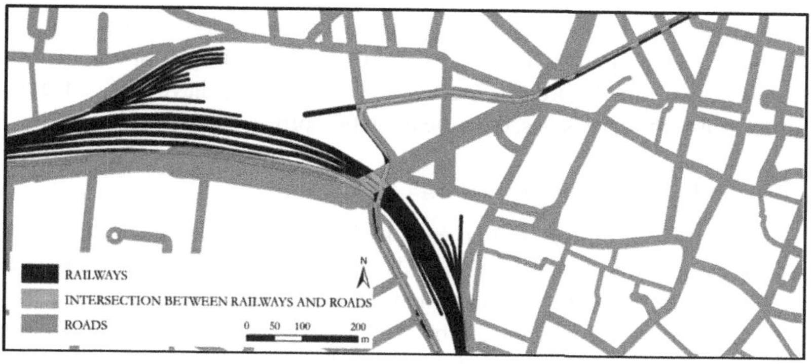

Figure 3.4. *Geometric overlaps between roads and railways. For a color version of the figure, see www.iste.co.uk/baghdadi/qgis3.zip*

QGIS functionality to remove layer overlaps:
- Remove overlapping layers: Vector > Geoprocessing Tools > Difference

3.2.2.2. Generation of buildings' footprint layer

Buildings' footprint can be defined either by:

– **According to the BD TOPO® footprint of buildings** (without change) from the BD TOPO® database. This surface is calculated based on the contours of the buildings as they appear from the sky (see Figure 3.5). Note that inner courtyards are included only if their width is greater than 10 m.

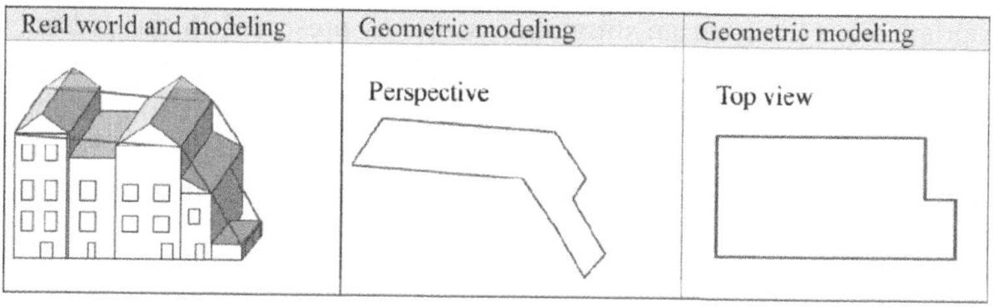

Figure 3.5. *Buildings' geometric modeling in BD TOPO® (source: [IGN 16])*

– **Or according to the generation of buffers around the buildings** based on methods currently used for urban fabric studies [CER 08, CER 13]. For example, the buffer around a building located in an industrial area includes the car park areas.

Usually, the buffer generation around buildings is used in the framework of continuous urban fabric studies in which the approach is based on a “dilatation-erosion” process [CER 08]. The dilatation process is inherently used to create a buffer around buildings whereas the erosion process ensures the reduction and the spatial assembling of buffers around buildings [CER 08]. Appropriate for interurban studies, the “dilatation-erosion” process seems to remove open space in the city (inner courtyards, small gardens, etc.). Consequently, we have defined a buffer generation around buildings based on the dilatation process alone.

3.2.2.2.1. Buffer definition for buildings

The application of buffers around each building aims to approach their physical characteristics in order to restore their “real” footprint by taking into account artificial areas that could not be identified based on the data provided by the BD TOPO® database (terraces, garages, outbuildings, etc.). Note that this processing does not allow us to measure the urban space expansion. The buffer definition for buildings requires three preliminary steps:

1) calculate the buildings’ surfaces;

2) select and remove the smallest buildings (< 20 m^2);

3) define a buffer around each building.

QGIS functionalities* to select and remove building features:
- Open attribute table > Open Select features using an expression window ...
- Select by expression: QGIS geoalgorithms > Vector Selection tools > Select by expression

The buffer definition may vary with land use and the targeted spatial scale. An oversized buffer will tend to reduce the open space in the city, which is inappropriate for a study at an intraurban scale, as presented in the present work. Conversely, if the buffer is undersized, little additional information will be provided about the spatial extent of buildings. In other words, the results strongly depend on the buffer radius chosen and on the targeted scale of study. In the CERTU methodology user guide [11] [CER 08], the buffer radius for an interurban approach typically ranges from 15 to 40 m.

For the Dijon-city case, different buffer radiuses were tested in order to choose the threshold, which allows open space to be observed in a certain district. The buffer radius is finally fixed at 10 m around each building. This makes it possible to

11 CERTU is a French institute center on Networks, Transport infrastructures, Urban Planning and Public Buildings. Since 2014, it has been integrated into a larger public authority called CEREMA.

observe the open spaces within a district area, which can be associated with the presence of individual gardens or inner courtyards as illustrated in the city center of Dijon (see Figure 3.6).

Nevertheless, although the buildings' footprint appears to be adequate for the city center and the residential areas, it will be underestimated in tertiary activity zones. Indeed these urban spaces have specific characteristics such as the presence of large car park areas around each building[12] and/or the presence of relatively large storage areas mainly constituted of artificial areas. Therefore, the buffer radius has been fixed to 15 m for buildings located in tertiary activity zones (see section 3.2.2.2.2 for details). In other words, two buffer radiuses were applied, depending on the land use, in order to account for the specificity of all buildings.

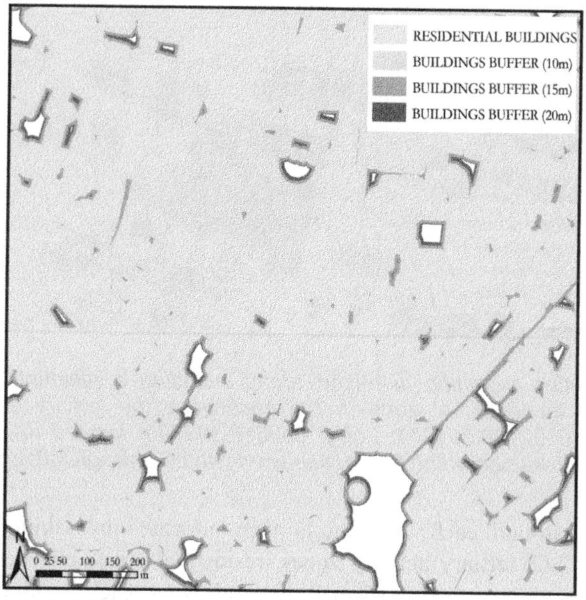

Figure 3.6. *Various buffer radius tests (10, 15 and 20 m) in dense building area: City center of Dijon. For a color version of the figure, see www.iste.co.uk/baghdadi/qgis3.zip*

3.2.2.2.2. Specific processing for buildings located in tertiary activity zones

First of all, the CLC database is used (raster file at a spatial resolution of 100 m) to identify the buildings located in tertiary activity zones. Then the combination

12 Note in the BD TOPO® database park areas less than ½ ha are not entered.

between the CLC raster layer and the buildings vector layer allows us to identify the land use for each building in the attribute table:

QGIS functionality to cross raster and vector layers:
- *Raster to target vector field > Zonal statistics > Zonal statistics...*

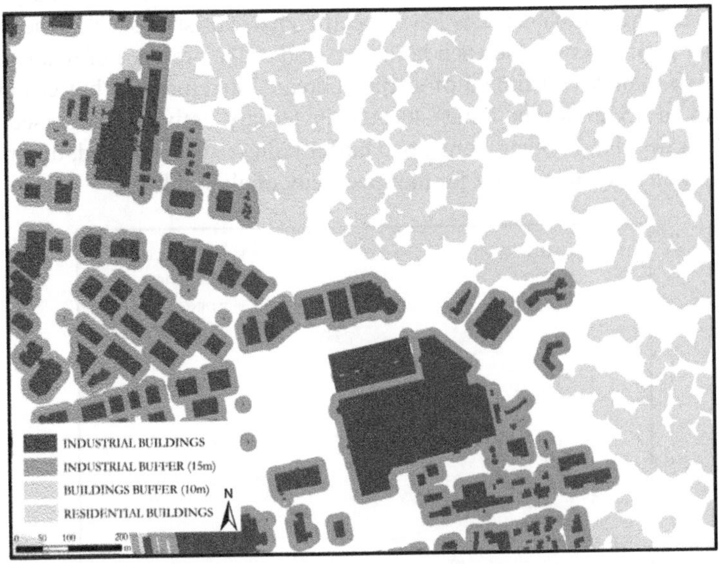

Figure 3.7. *Buffer allocation in mixed areas including a residential zone in the northeast, and an industrial zone in the southwest: the tertiary activity zone of Quetigny (Lat: 6692129.501971; Long: 858612.392790) located in the suburbs of Dijon. For a color version of the figure, see www.iste.co.uk/baghdadi/qgis3.zip*

The buffer around each building is then allocated by taking the land use (according to CLC: tertiary activity zones, residential areas, city center, etc.) into account. Hence, we assigned a 15 m buffer radius for each building located in a tertiary activity zone (CLC code: 3 and 4) and a 10 m buffer radius for all the other buildings. Both specific buffers are presented in Figure 3.7.

QGIS functionality to create a buffer from a new field in the attribute table:
- Open attribute table > Open the *field calculator (Ctrl-I)*

3.2.2.3. *Pre-processing of the vegetation areas*

The BD TOPO® database allows us to distinguish all the wooded and forest areas that have a surface greater than 500 m^2 [IGN 16]. As a result, the BD TOPO® is not appropriate for observing the urban vegetation diversity, for instance, herbaceous vegetation, which is a very frequent vegetation type in cities (and can take up a significant part of urban spaces). As shown in Figure 3.8, complementary information can be provided by the OSM database.

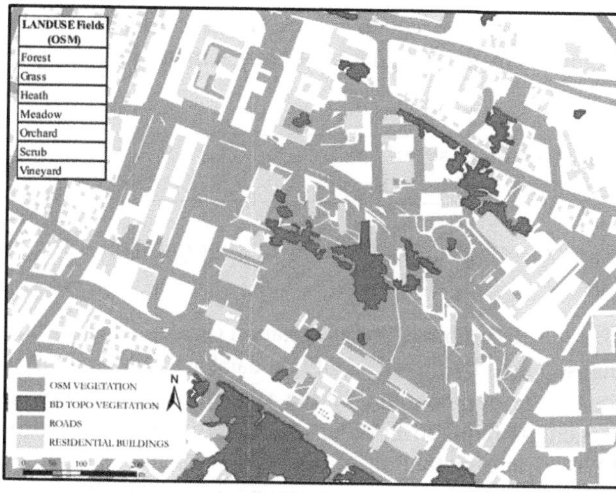

Figure 3.8. *Complementarity between vegetation from OpenStreetMap and BD TOPO® in open landscape: the University campus in Dijon. For a color version of the figure, see www.iste.co.uk/baghdadi/qgis3.zip*

Since the OSM database is based on a Volunteered Geographical Information (VGI) [GOO 07], the data provided has to be handled with caution. However, "*contrary to the* a priori *that one can have on the VGI data, accuracy remains respectable for a GIS use, especially in France*" [LAM 13]. In our case, the LANDUSE feature classes provided by the OSM database enable us to distinguish between different types of spaces that are considered vegetated. The fields to select and extract from the LANDUSE layer are detailed in Figure 3.8. As a result, two vegetation layers are available: one from the OSM database and the other one from the BD TOPO(r) database.

There are some geometric data overlaps, such as, for instance, the herbaceous surfaces of the OSM (grass class) and the forest areas (forest class) of the BD TOPO® (Figure 3.8). Thus there is a need to ensure the non-overlaps of the different vegetation areas before merging both vegetation layers (the one from the BD TOPO® and the other one from the OSM) as shown in Figure 3.8.

3.2.2.4. *Buffer definition and merging layers: generation of anthropization data*

The last step of the land use data processing chain is to merge, on the one hand, the areas identified and processed as "artificial" surfaces (buildings and transport infrastructures) and, on the other hand, the areas identified as "natural" surfaces (vegetation and hydrographic areas). For convenience, one file per surface type (artificial and natural) is generated in order to estimate the percentages of anthropization of the territory (step 3, see Figure 3.1). The geographic data processing is facilitated by the merging of all the geometries as a single spatial entity (dissolve):

QGIS functionality to merge layer:
- *Vector > Vector geometry tools > Dissolve*

3.2.2.4.1. Pre-processing of natural areas: vegetation and hydrographic areas

The natural area file combines both vegetation (presented in section 3.2.2.3) and water body data, supplied by the BD TOPO® database. Once again, it is important to pay attention to the non-overlaps of the vegetation and hydrological areas. A geometric processing is required in order to preserve the native hydrographic areas of the BD TOPO®, intersecting them with the vegetation areas' geometry (see section 3.2.2.1.3). Once the clipping processing is completed, the two layers are merged in a natural areas file that combines both vegetation and hydrographic areas.

3.2.2.4.2. Pre-processing of artificial areas: buildings' and transport infrastructures' areas

The pre-processing of artificial areas is similar to that of natural areas. The data allowing the generation of the artificial areas file include the footprints of the transport infrastructures (see section 3.2.2.1) and of the buildings (see section 3.2.2.2). The surface data related to the presence of car park and/or large motorway tolls areas are also integrated. Initially supplied by the BD TOPO® database (Thematic: SURFACE_ROADS), these geometric data allow all the road surfaces of more than 0.5 ha to be considered [IGN 16]. In order to easily manipulate the footprint data, it is appropriate to merge the artificial and natural layers in the same layer. However, before this operation, it is necessary to ensure the absence of overlaps of the information using the following GIS processing, performed step by step:

– **a joint processing of intersections between buildings and road surfaces**, which are then merged as surface data associated with buildings.

– **a joint processing of the previous operations related to the transport infrastructures** by a differentiation processing between the buildings and the transport infrastructures' surfaces.

Once the overlaps between the different surfaces are processed, the data merging enables us to generate the artificial areas file. A final phase of differentiation between natural and artificial areas is then conducted. This allows us to generate the information related to the anthropization of the territory in the vectorial format (Figure 3.9).

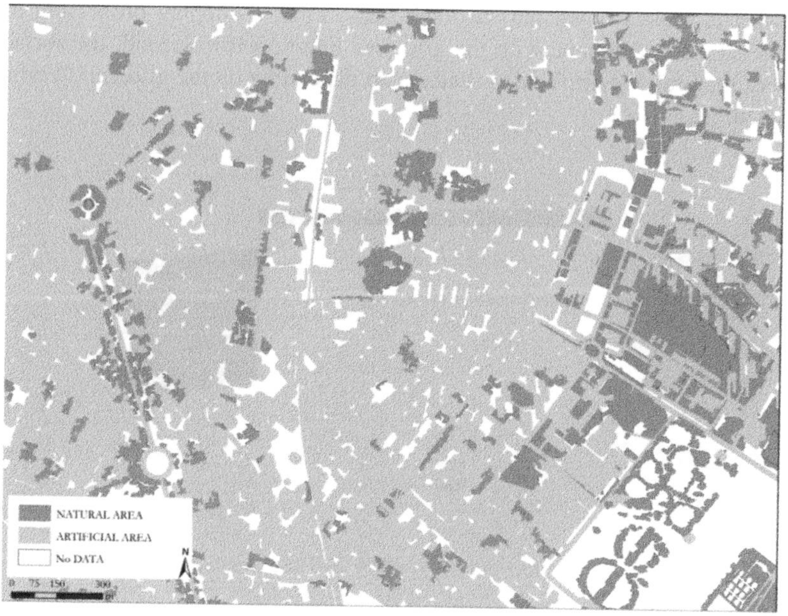

Figure 3.9. *Anthropization map at the end of step 2. For a color version of the figure, see www.iste.co.uk/baghdadi/qgis3.zip*

3.2.3. *Step 3: calculate the anthropization index*

The anthropization index is calculated in order to have a general view of the anthropization at the scale of the territory. This index is calculated by means of a regular grid, the spatial resolution of which is of high importance. In our case, the resolution is the one chosen for the MAM (resolution at 150 m × 150 m). In practice, this grid will be used as an intermediate layer (mask layer) in which three calculations will be carried out: the percentage of anthropization, the surface of natural areas and the surface of artificial zones[13]:

QGIS functionality* to create a regular grid:
- Generate a grid: *Vector > QGIS geoalgorithms > Vector Grid...*

13 Furthermore, a great deal of information can be associated with each grid point, such as, the average elevation, the average height of buildings, shadow effects, etc. This information allows for the use of different input parameters in MAM.

The first phase consists of calculating the surface of the artificial areas within each grid point. The following processing is performed[14]:

1) **Calculate the surface of the different areas inside each grid point:** this processing is broken down into two sub-steps. First, a clipping processing of the surfaces data in accordance with the grid (see Figure 3.10). Second, the surface of each area (natural and artificial) is calculated depending on the grid cell to which it is attached.

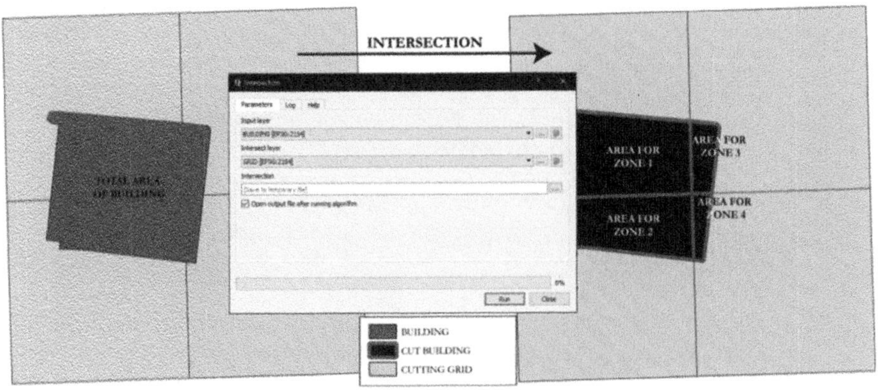

Figure 3.10. *Illustration of intersection GIS processing in accordance with a regular grid (QGIS functionality: Intersection). For a color version of the figure, see www.iste.co.uk/baghdadi/qgis3.zip*

> **QGIS functionality[*] to clip area in accordance with a grid:**
> - Cut areas with a polygon layer: *QGIS geoalgorithms > Vector Overlay tools > Intersection...*

2) **Calculate the percentage (index) of anthropization**: this last processing consists of a spatial join of the surface data of the grid. The information associated with each grid cell makes it possible to know whether the surfaces are artificial or natural within the grid cell. Note that white areas (spaces with no surface data) are considered natural spaces. The calculation of the percentage of anthropization for each grid cell is then carried out based on the following equation:

$$Percentage\ of\ anthropization_i = \frac{\text{Artificial areas}}{Total\ area_i} \times 100 \qquad [3.1]$$

14 Note that to reduce processing times it may be appropriate, but optional, to merge all surface geometries as single geometric data (QGIS functionality *dissolves*).

where $Percentage\ of\ anthropization_i$ stands for the surface of the artificial areas for each grid cell i and $Total\ area_i$ stands for the surface of each target grid of 150 × 150 m point (that is to say 22,500 m² / grid point i). The final result is an image of the anthropization of the territory at a resolution of 150 m, as illustrated in Figure 3.11.

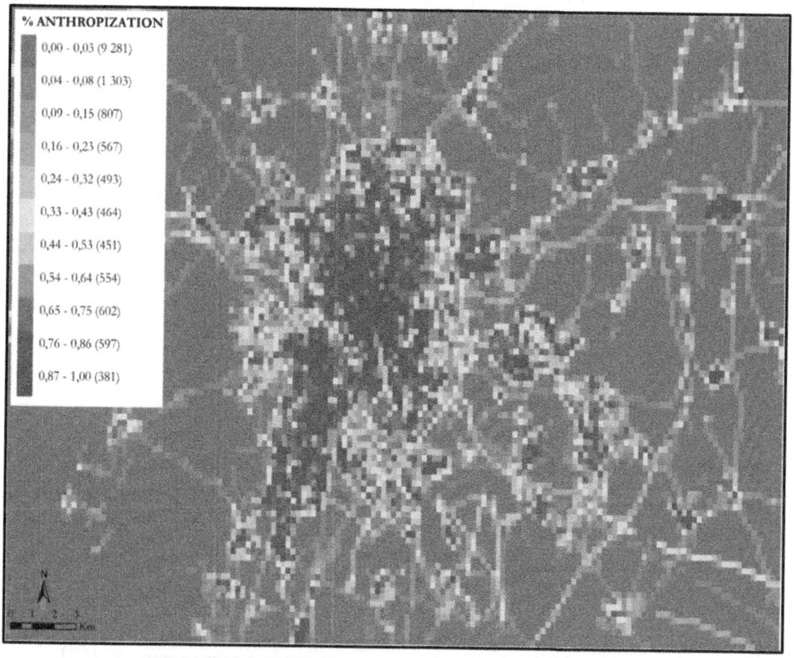

Figure 3.11. *Anthropization index at a resolution of 150 m in the study area (in brackets: the number of features associated at each % class). For a color version of the figure, see www.iste.co.uk/baghdadi/qgis3.zip*

3.2.4. Discussions and perspectives: contribution of remote sensing to vegetation mapping

The goal of this section is to complete the previous estimations using high resolution remote sensing products. If processings based on vectorial's datasets are centered on urban elements like buildings and transport infrastructures, processings associated with satellite imagery allow us to extract accurate information on the natural surfaces, especially vegetation through the use of vegetation indices like the NDVI (Normalized Difference Vegetation Index). This information may be of high interest in the frame of the present study as the bias related to the processings presented in this chapter are mainly due to vegetation (see section 3.2.2.3). This point is highlighted in Figure 3.11, where a strong disparity appears between urban and rural spaces. Here, it is interesting to note that the spaces located outside the urban areas are mainly natural (100%) whereas in fact these spaces are mixed, as suggested in Figure 3.11. In order to integrate the remotely sensed vegetation product in the existing processing chain, we have extracted the NDVI from the Landsat imagery (spatial resolution of 30 m) provided by the French Land Data Center Theia[15]. The NDVI are calculated based on surface reflectances (top of atmosphere [TOA] reflectances are corrected from atmospheric effects), in the red (ρ_{Red}) and the near infrared (ρ_{NIR}):

$$NDVI = \frac{\rho_{NIR} - \rho_{Red}}{\rho_{NIR} + \rho_{Red}} \quad [3.2]$$

The NDVI theoretically ranges between –1 and 1, negative values being attributed to water bodies. According to [SOB 04], the classification of NDVI allows three vegetation types within an urban space to be defined:

– class 1: the artificial areas: $0 < \text{NDVI} < 0.20$;

– class 2: the mixed areas: $0.20 \leq \text{NDVI} \leq 0.5$;

– class 3: the vegetation areas: $\text{NDVI} > 0.5$.

In the present study, only areas dominated by vegetation (class 3, NDVI > 0.5) are taken into account. For each cell *i* of the targeted grid (spatial resolution of 150 m), it is possible to calculate the number of pixels_k (30 m × 30 m) attached to each vegetation class. The percentage of vegetation in each grid cell *i* is then calculated as follows:

$$Percentage\ of\ vegetation_i = \frac{\text{Number of pixel}_k\text{ with NDVI>0.5}}{\text{Total number of Pixels}_k} \times 100 \quad [3.3]$$

Finally, the combination between the two indicators (anthropization through GIS processings and vegetation through remote sensing) is illustrated in Figure 3.12. Note that the final map is at a spatial resolution of 30 m.

The urban areas located outside of the city of Dijon are more heterogeneous in Figure 3.12 than in Figure 3.11. This heterogeneity highlights a strong contrast between the forest type and the mixed areas, which is quite specific to the agricultural areas. Focusing now on Dijon, urban parks can be observed within densely built areas. The latter result illustrates the complementarity between GIS processing (for artificial areas) and remote sensing processing (for vegetated areas) very well. As a result, remote sensing products at high (or very high) spatial resolution may help to refine the anthropization index calculation by better

15 Landsat data by THEIA: https://theia-landsat.cnes.fr/rocket/#/search?collection=Landsat.

highlighting the contrast between artificial and natural areas, outside the urban areas (for instance, forest versus urban areas) or within the city center (for instance parks within densely built areas).

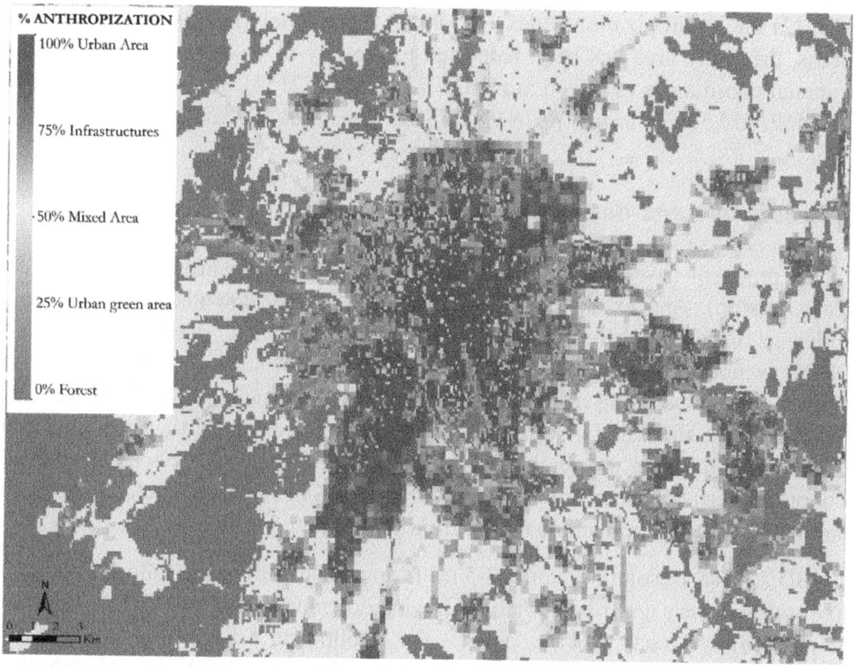

Figure 3.12. *Anthropization index of the territory at 30 m of spatial resolution. For a color version of the figure, see www.iste.co.uk/baghdadi/qgis3.zip*

To conclude, it is necessary to note that the size of a study domain as well as its spatial resolution are key determiners to choose the most adapted method. A processing performed on a relatively large domain of study will require a non-negligible calculation time. The latter will increase all the more if the processings are done at the scale of the study area. Our method seems more adapted to relatively restricted domains. Indeed, the transition between a 30 m and a 150 m spatial resolution inherently tends to provide a rather smooth and continuous morphology of the urban space. Finally, in this chapter, we illustrate the complementarity between two products of different types: on the one hand, vectorial data processed at 150 m of spatial resolution (BD TOPO®) and, on the other hand, remote sensing data provided at 30 m of spatial resolution (LANDSAT images). As a perspective, we propose to evaluate these results at a finer scale based on very high resolution remote sensing products (for instance based on Pléiades images).

3.3. Practical case study of the processing chain

This section presents a practical case study of the processing chain previously described. The goal is to obtain: (1) the integration of the buildings' heights into a DTM; (2) the anthropization index. To obtain more timely results about the buffer process, the study area has been restricted to the city of Quetigny (East of Dijon), the spatial coordinates of which are (RGF 93 - EPSG: 2154): *xmin: 857340.573349; ymin: 6691455.57952; xmax: 861840.573349; ymax: 6695205.57952.*

3.3.1. *Software and database*

This practical case was conducted using the QGIS software (2.14[16] long-term release version). Any extensions needed are additionally downloaded. This choice enables us to apply different processings using the same default version. BD TOPO® is not an open access database. The practical study presented here was therefore elaborated using a sample database provided by the OSM. However, it is advisable to use the BD TOPO® database or an other database that provides a 3D description of building morphologies[17]. This third dimension allows us to take into account the vertical dimension of the city like the SVF. In this practical study, the input datasets are downloadable online from the following sources:

– BD ALTI® (resolution 75 m × 75 m) is downloaded from the following address: http://professionnels.ign.fr/bdalti#tab-3. If you download the BD ALTI® database, you should extract DTM from the following folder: ~\BDALTIV2_2-0_75M_ASC_LAMB93-IGN69_FRANCE_2017-01-04\BDALTIV2\1_ DONNEES_ LIVRAISON_2017-02-00100\BDALTIV2_DTM_75M_ASC_LAMB93_IGN69_ FRANCE**BDALTIV2_75M_FXX_0825_6750_DTM_LAMB93_IGN69.asc,** which will be renamed as "DTM_BOURGOGNE".

– OSM databases are available from: http://download.geofabrik.de/ europe/france/ bourgogne.html. The folder name that should be downloaded is: bourgogne-latest-free.shp.zip which will be renamed as "BOURGOGNE_ OSM".

– The CLC dataset of *Côte d'Or* at a resolution of 25 m is downloaded here for the 2006 year (revised version: "CLC06R_D021_RGF.tif"): http://www. statistiques.developpement-durable.gouv.fr/clc/fichiers/.

16 Web link to QGIS software: http://qgis.org/en/site/forusers/download.html.

17 In this practical case study building height is not available, thus we used a fictitious height.

3.3.2. *Step 1: integrate the building heights into a DTM*

3.3.2.1. *BD ALTI® data pre-processing*

Approach	QGIS handling
1. Convert ASC format to Tiff **OPTIONAL**	In the QGIS menu bar: • Click on **Raster > Conversion > Translate (Convert Format)…** In translate tool: • Select the input layer: BDALTIV2_75M_FXX_0825_6750_DTM_LAMB93_IGN69.asc • Set the output filename: DTM_BOURGOGNE • Select the target CRS: RGF93/Lambert-93/EPSG 2154
2. Clip the study area	In the QGIS menu bar: • Click on: **Raster > Extraction > Clipper…** In Clip tool: • Select the input layer: DTM_BOURGOGNE • Set the output filename: DTM_DOM • Click on extend mode and select this extent coordinate: xmin: 857340.573349; ymin: 6691455.57952; xmax: 861840.573349; ymax: 6695205.57952

Table 3.3. *Pre-processing steps of the DTM*

3.3.2.2. *Buildings data pre-processing*

Approach	QGIS handling
1a. Reproject and clip the building layer	In QGIS: • Open the following OSM shapefile: gis.osm_buildings_a_free_1 In the QGIS Processing Toolbox: • Click on QGIS geoalgorithms> Vector general tools> Reproject layer • Select the target CRS: RGF93/Lambert-93/EPSG 2154 In the QGIS Processing Toolbox: • Click on GDAL/OGR > [OGR] **Geoprocessing > Clip vectors by extend**

	In Clip tool: • Input layer • Select the following clip extend: *xmin :857340.573349; ymin :6691455.57952; xmax :861840.573349; ymax :6695205.57952* Set the output filename: BUILDING
1b. Create a buildings height features into a new field (if OSM database is in use) **OPTIONAL**	Open the "BUILDING" attribute table: • Open the field calculators (Ctrl+I) In the field calculators: • Create a new field "HEIGHT" • Define output field length as 4 • Enter the following expression: Rand(1,10): random attribute layer between 1 and 10 • Save edits • Close editing mode (Ctrl+E)
2. Convert a shapefile to raster	In the QGIS menu bar: • Click on **Raster** > **Conversion** > **Rasterize (Vector to Raster)...** In rasterize tool: • Select the input layer: BUILDING • Select the attribute field: HEIGHT • Set the output filename: BUILDING • Define the raster resolution in map units per pixel at 1 to 1 m

Table 3.4. *Rasterization of buildings' height*

3.3.2.3. *DTM and building data integration*

Approach	QGIS handling
1. Resample DTM raster data to a finer grid (interpolation)	In the QGIS menu bar: • Click on **Processing** > **Toolbox (keyboard shortcut: Ctrl+Alt+T)** In Toolbox: • Open the interpolation resample tool: **GRASS commands** > **Raster** > **r.resamp.interp**

	In resample interpolation tool: • Select the input layer: DTM_DOM • Select the sampling interpolation method: bilinear • Select the DTM_DOM as region extent with the following option (**… > Use layer/canvas extend** >): 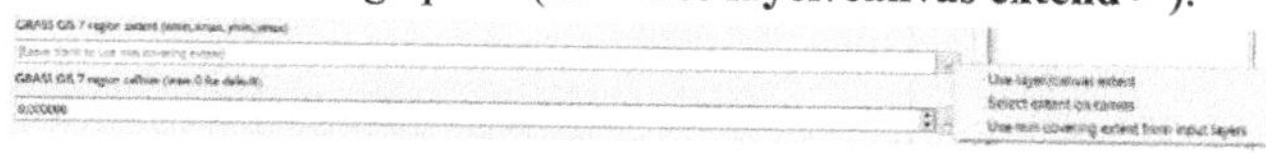 • Define a region cellsize at 5 m • Set the output filename: DTM_DOM_5m
2. Resample the buildings' heights raster data to a coarser grid (aggregation)	In the QGIS menu bar: • Click on **Processing** > **Toolbox (keyboard shortcut: Ctrl+Alt+T)** In Toolbox: • Open the aggregation resample tool: **GRASS commands** > **Raster** > **r.resamp.stats** In resample aggregation tool: • Select the input layer: BUILDING • Select aggregation method: mode • Select the DTM_DOM as region extent with the following option (**… > Use layer/canvas extend** >) • Define a region cellsize at 5 m • Set the output filename: BUILDING _5m
3. Merge raster	Open the **Raster Calculator (Raster Toolbar** > **Raster Calculator…)**: • Enter the following expression: "BUILDING_ 5m@1" + "DTM_DOM_5m@1" • Select the previous region extent • Set the output filename (GeoTIFF): DTM_BUILDING

Table 3.5. *Integration steps of buildings' height to DTM*

3.3.3. *Step 2: geographic data generation about natural and artificial areas*

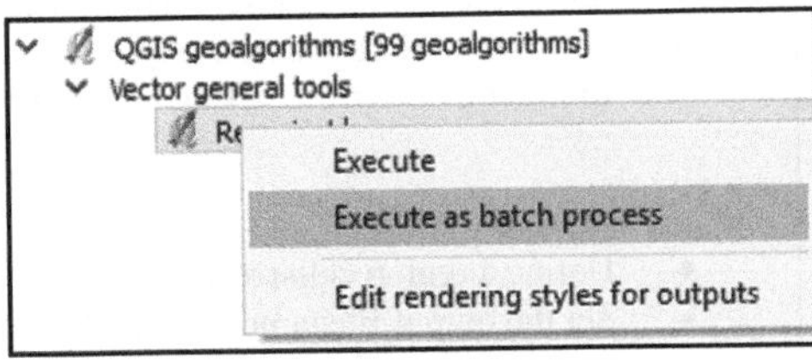

This step aims to process OSM vectorial layers related to transport infrastructures, buildings, vegetation and hydrographic areas. In order to optimize the calculation time, a batch processing tool is highly recommended to be used in order to reproject and clip the datasets. The batch mode is available by: (1) clicking on "Run as batch process…" as shown below, or by (2) right clicking on the Processing Toolbox and selecting the "Execute as batch process" as shown):

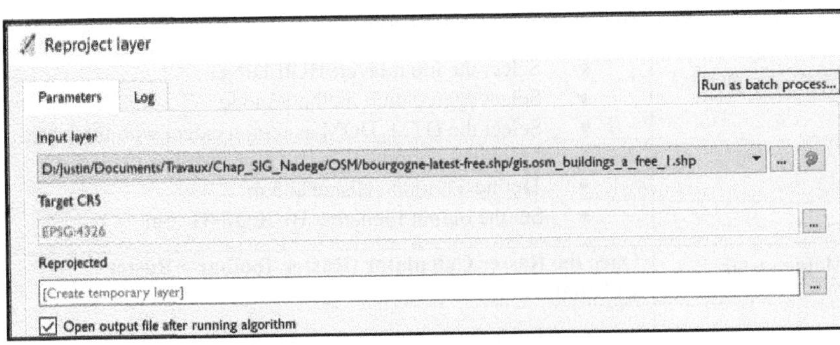

Figure 3.13. *Batch process with QGIS*

The layers reprojections in batch mode; Click on "**QGIS geoalgorithms > vector general tools > reproject layer**". The WGS84 native project (CRS: EPSG: 4326) is then converted into Lambert RGF93 (Target CRS EPSG: 2154), the output layers are renamed as follows:

– gis.osm_landuse_a_free_1 renamed LANDUSE

– gis.osm_railways_free_1 renamed RAILWAYS

– gis.osm_roads_free_1 renamed ROADS

– gis.osm_water_a_free_1 renamed WATER_AREA

– gis.osm_waterways_free_1 renamed WATER_WAY

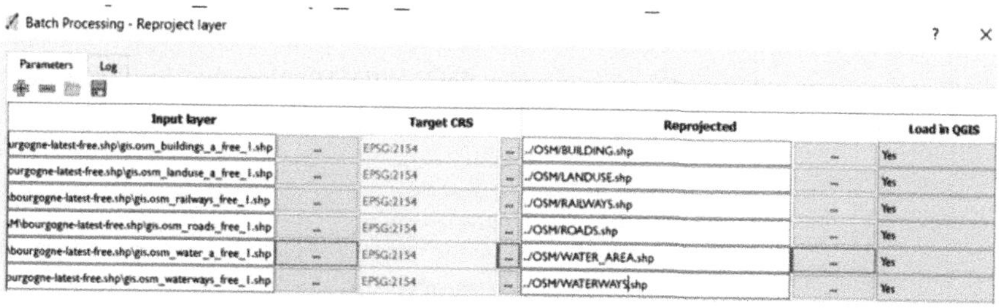

Figure 3.14. *Reprojection of layers in batch process mode*

Then to clip the study area domain click on "**GDAL/OGR > Geoprocessing > Découper**", select the following extent coordinate: *xmin: 857340.573349; ymin: 6691455.57952; xmax: 861840.573349; ymax: 6695205.57952*. Afterward, rename the output layers as follows:

– BUILDING renamed BUILDING_DOM

– LANDUSE renamed LANDUSE_DOM

– RAILWAYS renamed RAILWAYS_DOM

– ROADS renamed ROADS_DOM

– WATER_AREA renamed WATER_AREA_ DOM

– WATER_WAY renamed WATER_WAY_ DOM

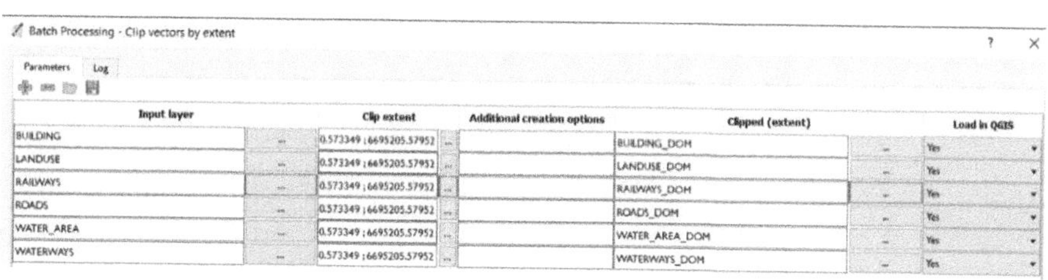

Figure 3.15. *Clipping of layers in batch process mode*

3.3.3.1. *Transport infrastructures pre-processing*

Approach	QGIS handling
1. Define the railway widths	Input vector layer: RAILWAYS_DOM Open the attribute table: • Open the field calculators (Ctrl+I)

	In the field calculators: • Create a new field "WIDTH". Note that this should be a defined decimal number (real) with precision 3 at least: 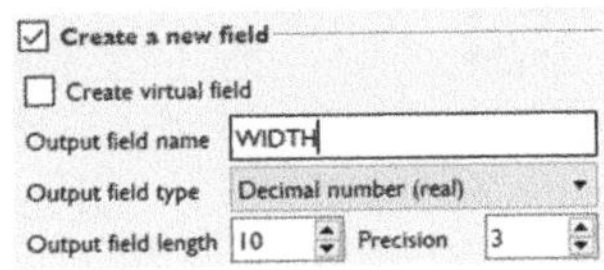 • Enter the following expression: 3.935 • Save edits • Close editing mode (Ctrl+E)
2. Create the railway area	In the QGIS menu bar: • Click on **Vector > Geoprocessing Tools> Buffer…** In the variable distance buffer tool: • Select as input layer: RAILWAYS_DOM • Select as Distance field: WIDTH • Click on Dissolve result • Set the output filename: RAILWAYS_AREA
3. Define the road widths[18]	Input vector layer: ROADS_DOM Open the attribute table: • Open the field calculators (Ctrl+I) In the **field calculators**: • Create a new field "WIDTH". • Enter the following expression: CASE WHEN "fclass"= 'cycleway' THEN 2.5 WHEN "fclass"='motorway' THEN 20 WHEN "fclass"='motorway_link' THEN 4 WHEN "fclass"='primary' THEN 10 WHEN "fclass"='primary_link' THEN 2.5 WHEN "fclass"='residential' THEN 5 WHEN "fclass"='secondary' THEN 8 WHEN "fclass"='secondary_link' THEN 2.5

18 Unlike the BD TOPO® database, OSM database does not integrate information about the number and the width of road lanes. Therefore, based on the BD TOPO® roads database, we have defined the average values of road widths.

	WHEN "fclass"='tertiary' THEN 5 WHEN "fclass"='tertiary_link'THEN 2.5 WHEN "fclass"='trunk' THEN 18 WHEN "fclass"='trunk_link' THEN 4 ELSE 0 END • Save edits • Close editing mode
4. Create the roads area	In the QGIS menu bar: • Click on **Vector** > **Geoprocessing Tools> Buffer…** In variable distance buffer tool: • Select as input layer: ROADS • Select as Distance field: WIDTH • Click on Dissolve result • Set the output filename: ROADS_AREA[19]
5. Remove geometrics overlapping layers: Difference process	In QGIS processing Toolbox: • Click on **SAGA** > **Vector polygon tools** > **Difference…** In Difference tool: • Select as layer A: RAILWAYS_AREA • Select as layer B: ROADS_AREA • Set the output filename: RAILWAYS_AREA2
6. Create transport infrastructures areas	In the QGIS menu bar: • Click on **Vector** > **Data management tools** > **Merge vector layers** In Merge vector layers tool: • Select the previous layers to merge: RAILWAYS_AREA2 and ROADS_AREA • Set the output filename: INFRASTRUCTURES

Table 3.6. *Creation steps of transport infrastructures areas*

19 Note that some routes do not appear in the buffer because the given road width is equal to 0.

3.3.3.2. *Building data pre-processing*

Approach	QGIS handling

1. Calculate area of the buildings and select feature from attribute table	• Input vector layer BUILDING
	Open the attribute table and the field calculators
	In the field calculators:
	• Create a new field: AREA
	• Enter the following expression:
	`$area`
	• Open the Select by expression window:

	• Enter the following expression:
	`"AREA" <= 20`
	• Delete selected features
	• Save edits
	• Close editing mode

2. Assignment of building land uses from Corine Land Cover data	• Input raster layer CLC06R_D021_RGF.tif • Input vector layer BUILDING In the QGIS menu bar: • Click on **Raster** > **Zonal statistics** > **Zonal statistics …** In Zonal statistics tool: • Select the raster layer: CLC06R_D021_RGF • Select polygon layer containing zones : BUILDING • Set a output column predix: "CLC_" • Select statistics to calculate: majority
3. Calculate building footprints according to the type of land use	Open the attribute table from BUILDING layer In the field calculators: • Create a new field: WIDTH • Enter the following expression: CASE WHEN "CLC_majori" = 121 OR "CLC_Majori"= 122 THEN 15 ELSE 10 END

	• Save edits • Close editing mode
4. Create building footprints	In the QGis menu bar: • Click on **Vector > Geoprocessing Tools > Buffer…** In variable distance buffer tool: • Select as input layer: BUILDING • Select as Distance field: WIDTH • Click on Dissolve result • Set the output filename: BUILDING_AREA

Table 3.7. *Creation steps of buildings' footprints*

3.3.3.3. *Vegetation and hydrographic data pre-processing*

Approach	QGIS handling
1. Extract vegetation areas from LANDUSE layer	• Input vector layer LANDUSE Open the attribute table LANDUSE layer • Open the Select by expression window:  • Enter the following expression: "fclass" IN ('forest','grass','heath','meadow','orchard', 'Scrub','vineyard') • Close the window Save only selected features: • Right click on LANDUSE layer • Save as… • Check the box for "Save only selected features" • Set the output filename: VEGETATION

	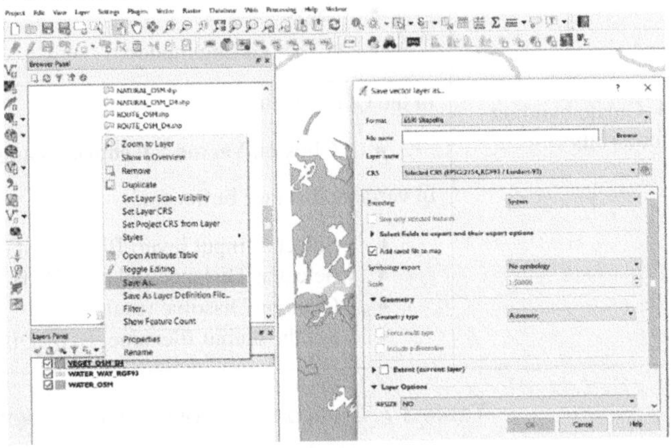
2. Create hydrographic areas	• Input layer WATER_WAY In the QGIS menu bar: • Click on **Vector > Geoprocessing Tools > Buffer…** In Fixed distance buffer tool: • Select as input layer: WATER_WAY • Select a buffer distance to 8 m • Click on Dissolve result • Set the output filename: WATER_WAY_AREA In QGIS Processing toolbox: • Click on **SAGA > Vector polygon tools > Difference…** In Difference tool: • Select as layer A: WATER_WAY_AREA • Select as layer B: WATER_AREA • Set the output filename (overwrite the previous WATER_WAY_AREA layer): WATER_WAY_AREA In the QGIS menu bar: • Click on **Vector > Data management tools > Merge vector layers** In Merge vector layers tool: • Select the following layers to merge: WATER_WAY_AREA and WATER_AREA • Set the output filename: HYDROGRAPHY

Table 3.8. *Pre-processing steps of natural areas*

3.3.3.4. *Create natural areas layer*

Approach	QGIS handling
1. Create the natural areas layer	• Input layers: VEGETATION and HYDROGRAPHY In QGIS Processing toolbox: • Click on **SAGA > Vector polygon tools > Difference…** In Difference tool: • Select as layer A: HYDROGRAPHY • Select as layer B: VEGETATION Set the output filename (overwrite the previous HYDROGRAPHY layer): HYDROGRAPHY In QGIS Processing toolbox: • Click on **SAGA / Vector general tools / Merge vector layers…** In Merge vector layers tool: • Select input layers: HYDROGRAPHY and VEGETATION • Set the output merged layer: NATURAL_AREA

Table 3.9. *Creation steps of natural areas*

3.3.3.5. *Create artificial areas layer*

Approach	QGIS handling
1. Remove geometrics overlapping layers between infrastructures and building areas	In QGIS Processing toolbox: • Click on **SAGA > Vector polygon tools > Difference…** In Difference tool: • Select as layer A: BUILDING_AREA • Select as layer B: INFRASTRUCTURES Set the output filename: BUILDING_FINAL
2. Create the artificial areas layer	In QGIS Processing toolbox: • Click on **SAGA / Vector general tools / Merge vector layers…** In Merge vector layers tool: • Select input layers: BUILDING_FINAL and INFRASTRUCTURES • Set the output merged layer: ARTIF_AREA_tmp

3. Create the final artificial areas layer	In QGIS Processing toolbox: • Click on **SAGA > Vector polygon tools > Difference…** In Difference tool: • Select as layer A: ARTIF_AREA_tmp • Select as layer B: NATURAL_AREA Set the output filename: ARTIF_AREA

Table 3.10. *Creation steps of artificial areas*

3.3.4. *Step 3: calculation of the anthropization index*

Approach	QGIS handling
1. Create a grid layer	In QGIS Processing toolbox: • Click on **QGIS geoalgorithms > Vector creation tools > Create grid** In create grid tool: • Select the grid type: rectangle (polygon) • Select the following clip extend: *xmin :857340.573349 ; ymin :6691455.57952 ; xmax :861840.573349 ; ymax :6695205.57952* • Set a horizontal and vertical spacing to 150 m • Select as CRS Grid: 2154 (RGF 93 – Lambert 93) Set the output filename: GRID_ANTHROP
2. Create a unique ID in a new field	Open the attribute table: • Open the field calculators (Ctrl+I) In the field calculators: • Create a new field “ID” • Enter the following expression: @row_number • Save edits • Close editing mode
3. Intersection between area layer and grid	In QGIS Processing toolbox: • Click on **SAGA > Vector polygon tools > Intersect** • Select as layer A: GRID_ANTHROP • Select as layer B: ARTIF_AREA • Set the output filename: GRID_ARTIF_tmp

<table>
<tr>
<td></td>
<td>
In QGIS Processing toolbox:
<ul>
<li>Click on SAGA > Vector polygon tools > Polygon dissolve (by attribute)</li>
<li>Select as Polygons layer: GRID_ARTIF_tmp</li>
<li>Select as attribute field: ID</li>
<li>Set the output filename (Dissolved): GRID_ARTIF</li>
</ul>
</td>
</tr>
<tr>
<td>4. Calculate surface areas</td>
<td>
Input layer: GRID_ARTIF

Open the attribute table and the field calculators

In the field calculators:
<ul>
<li>Create a new field: SURF_ARTIF</li>
<li>Enter the following expression:</li>
</ul>
<hr>
$area
<hr>
<ul>
<li>Save edits</li>
<li>Close editing mode</li>
</ul>
</td>
</tr>
<tr>
<td>5. Join attribute with surface area data</td>
<td>
Open the layer properties of GRID_ANTHROP layer

In the properties windows:
<ul>
<li>Click on Joins:</li>
<li>Add vector join:</li>
</ul>
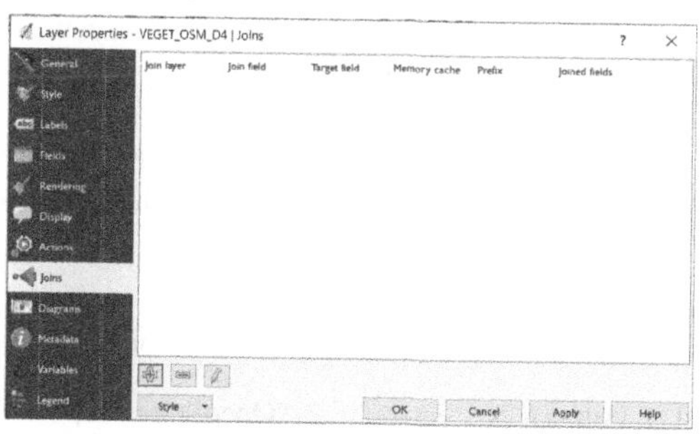

<ul>
<li>Join layer: GRID_ARTIF</li>
<li>Join field: ID</li>
<li>Target field: ID</li>
<li>Choose the following target field: SURF_ARTIF</li>
</ul>
</td>
</tr>
<tr>
<td>6. Calculate the anthropization index</td>
<td>
Input layer: GRID_ANTHROP

Open the attribute table and the field calculators

In the field calculators:
<ul>
<li>Create a new field: AREA</li>
<li>Enter the following expression:</li>
</ul>
<hr>
$area
<hr>
In the field calculators:
</td>
</tr>
</table>

	• Create a new field: ANTHROP • Enter the following expression: "GRID_ARTIF_SURF_ARTIF" / "AREA" Save the final layer as ANTHROP_FINAL.shp
7. Natural area treatment **OPTIONAL**	**For more information about vegetation areas, you should perform steps 3–6 on NATURAL_AREA**

Table 3.11. *Calculation of the anthropization index of the territory*

3.4. Bibliography

[BRU 05] BRUNET R., FERRAS R., THERY H., *Les Mots de la Géographie: Dictionnaire Critique*, 3rd ed., La Documentation française, Montpellier, Paris, 2005.

[CER 13] CERTU, "Analyse de l'occupation des sols: Constitution d'une tache urbaine", *Mesure de la consommation d'espace: méthodes et indicateurs,* Collection Dossiers, Fiche 1.3, Lyon, 2013.

[CER 08] CERTU, *Détermination d'un MOS et calcul d'une tache urbaine à partir de la BD TOPO de l'IGN*: Etude expérimentale, CERTU, Lyon, 2008.

[ESC 91] ESCOURROU G., *Le climat et la ville*, Nathan University, Paris, 1991.

[GOO 07] GOODCHILD M.F., "Citizens as sensors: the world of volunteered geography", GeoJournal 69, pp. 211–221, available at: https://doi.org/10.1007/s10708-007-9111-y, 2007.

[IGN 16] IGN, "Descriptif de la BD TOPO : Version 2.1.", available at http://professionnels.ign.fr/ sites/default/files/DC_BDTOPO_2_1.pdf, Saint Mandé, 2016.

[LAM 13] LAMBERT N., ZANIN C., "OpenStreetMap: collaborer pour faire des cartes", *M@ppemonde*, vol. 110, no. 2, available at: http://mappemonde-archive.mgm.fr/num35/ internet/int12301.html, 2013.

[LE 15] LE BRAS J., MASSON V., "A fast and spatialized urban weather generator for long-term urban studies at the city-scale", *Frontiers in Earth Science*, vol. 3, 2015.

[MAS 15] MASSON V., HIDALGO J., AMOSSÉ A. *et al.*, "Urban Climate, Human behavior & Energy consumption: from LCZ mapping to simulation and urban planning (the MapUCE project)", *9th International Conference on Urban Climate. CNRM - Météo France*, Toulouse, France, 2015.

[OKE 73] OKE T.R., "City size and the urban heat island", *Atmospheric Environment*, vol. 7, pp. 769–779, 1973.

[ROU 14] ROUX J., L'îlot de Chaleur Urbain à Dijon: Mise en place d'une campagne de mesures urbaines de la température, Master's thesis, University of Burgundy, Dijon, 2014.

[RYU 11] RYU Y.H., BAIK J.J., LEE S.H., "A new single-layer urban canopy model for use in mesoscale atmospheric models", *Journal of Applied Meteorology and Climatology*, vol. 50, pp. 1773–1794, 2011.

[SOB 04] SOBRINO J.A., JIMÉNEZ-MUÑOZ J.C., PAOLINI L., "Land surface temperature retrieval from LANDSAT TM 5", *Remote Sensing of Environment*, vol. 90, pp. 434–440, 2004.

[VEY 97] VEYRET Y., PECH P., *L'homme et l'environnement*, 2nd ed., Presses Universitaires de France, Paris, 1997.

4

Airborne Optical Remote Sensing Potential for Pool Mapping in an Urban Environment

4.1. Context

For a given urban scene, the pool map is defined as the knowledge in time of the position, boundaries and other information (negligence, volume, etc) of each pool.

This map has several advantages for urban managers:

– detection of neglected pools where dangerous mosquitoes can develop. Their bite on a human causes the West Nile virus [KIM 11];

– an accurate localization of water volumes. This water is usable by firemen in case of fire [ROD 14];

– a good consideration of the energy costs associated with the maintenance of pools [GAL 09];

– a control of declared estates by tax services [PHI 17].

Given the high number of properties with pools, it is not possible to carry out the mapping manually by field campaign or by photointerpretation on aerial images, for example. Indeed, this would not be exhaustive and would not allow for regular updating or to cover large scales. For example, France was the second largest market in the world for private pools behind the United States in 2010 with 1,465,840 pools

Chapter written by Josselin AVAL and Thierry ERUDEL.

at the end of 2009 compared with 708,000 in 2000 [LE 10], demonstrating a large number of pools and significant dynamics.

Remote sensing is "the set of techniques used to characterize objects remotely from the radiations they emit and/or reflect" [LIE 08]. In this definition, we understand that there are different remote sensing techniques, mainly depending on the distance and the radiation measured. In terms of distance, there is remote sensing *in situ*, by drone, plane or satellite. In addition, radiation can be measured in the visible and infrared (IR, optical), microwaves, etc. Depending on the applications, a certain remote sensing technique may be more appropriate than another.

Today, remote sensing makes it possible to map pools in an urban environment [KIM 11] and the objective of this chapter is to show the potential of airborne optical remote sensing for this mapping with QGIS software and Python scripts, which are free tools. The definition of the data required to study this potential is detailed in section 4.2.

4.2. Method

The main steps to study this potential are detailed below. Figure 4.1 illustrates the method:

1) data acquisition and preprocessing;

2) reference map definition;

3) feature extraction;

4) classification;

5) prediction map definition;

6) performance assessment.

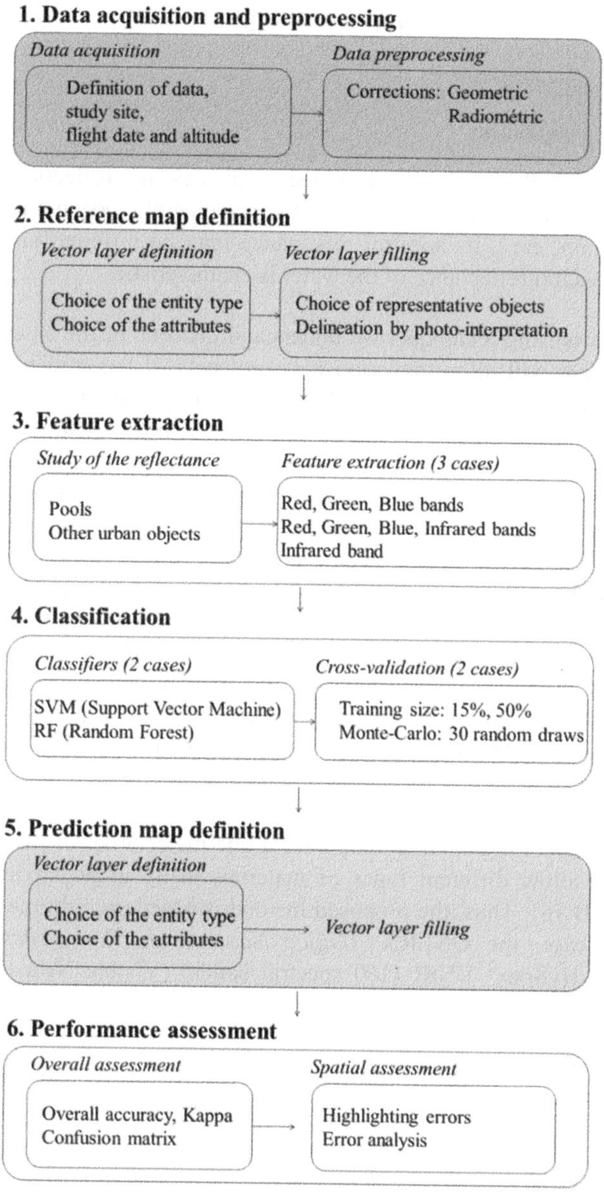

Figure 4.1. *Method for studying the potential of airborne optical remote sensing for pool mapping in an urban environment, For a color version of the figure, see www.iste.co.uk/baghdadi/qgis3.zip*

4.2.1. *Data acquisition and preprocessing*

4.2.1.1. *Data acquisition*

4.2.1.1.1. Data definition

The spectral reflectance of an object indicates its reflection property of electromagnetic radiation as a function of the wavelength. Depending on its nature (water, vegetation, etc.), its spectral reflectance may be different and we will see later that the spectral reflectance of the water is characteristic.

In a remote sensing technique, we understand that one of the essential elements is the imager that will allow radiation measurements of the studied objects to be acquired. Specifically, the hyperspectral imagers were designed to acquire an image in which each of the pixels contains a large number of measurements that correspond to particular wavelengths (spectral bands) of the electromagnetic radiation. This makes it possible to estimate the spectral reflectance of the objects of the urban scene with a high spectral resolution (also called sampling of the spectral reflectance). There are various hyperspectral imagers, e.g. Hyperion (220 spectral bands between 0.4 and 2.5 µm with a spatial resolution of 30 m), which is installed in the Earth Observing-1 (EO-1) and AVIRIS (Airborne Visible/Infrared Imaging Spectrometer) (224 spectral bands from 0.4 to 2.5 µm with a 1 m spatial resolution order of magnitude) satellites that are airborne. Both are developed by the National Aeronautics and Space Administration (NASA). In addition, the Environmental Mapping and Analysis Program is a German hyperspectral satellite whose launch is scheduled for 2019.

Today, airborne hyperspectral data have a one meter order of magnitude spatial resolution and allow different types of materials in an urban environment to be classified [AKB 16]. Thus, the proposed method is based on airborne hyperspectral data. In this case, the ONERA[1] (*Office National de Recherches et d'Études Aérospatiales*) HySpex[2] VNIR (160 spectral bands) (Visible Near-Infrared; 0.4-1 µm) camera is used.

QGIS functionality to open an image:
Layer > Add Raster Layer > ...

1 ONERA: French research center from which the DOTA department (*Département Optique et Techniques Associées*) acquires and operates remote sensing data.

2 HySpex: A Norwegian company that produces hyperspectral cameras.

4.2.1.1.2. Definition of study site, flight date and altitude

The study site must be representative of the urban environment so that the potential of airborne hyperspectral data for pool mapping, demonstrated for this case, can be generalized to the entire city or even to other cities.

The flight should not take place during the winter period because the pools are likely to be covered and therefore more complex to map because the spectral reflectance of the tarpaulins can be confused with that of other types of materials.

In order to distinguish pools, it is necessary to have a one meter order of magnitude spatial resolution. It is computed according to the altitude and the instrument parameters. The lower the altitude, the finer the spatial resolution.

4.2.1.2. *Data preprocessing*

4.2.1.2.1. Geometric preprocessing

During data acquisition, an airborne image undergoes a certain number of geometric deformations that can be grouped into two categories:

– the deformations related to the shot, caused by the instability of the aircraft (roll, pitch and yaw) or else by the angle of the shot. Knowledge of the operation of the sensor and taking into account the flight parameters make it possible to partially correct these errors;

– the deformations related to the curvature, the rotation of the Earth and to the relief, which are more complex to model.

The impact of the deformations related to the instability of the aircraft or to the rotation of the Earth can result in a blurred image. Special preprocessing operations are necessary to improve the sharpness of the images but they will not be discussed in this chapter: this effect is negligible in our case. Moreover, the deformations associated with the angle of inclination, the curvature of the Earth and the relief (so-called geometrical deformations) may cause the distance between two adjacent pixels to be different within the image. This may make it difficult to superimpose two images acquired in different configurations. This is analogous to someone who would observe a city from a tower: he would see the buildings near the tower larger than the buildings further away. In our case, these effects are sufficiently significant to be taken into account.

Correcting the geometrical deformations makes it possible to put the image in a reference frame, so that images acquired at different dates with potentially different imagers can be superimposed. From a modeling of these deformations, the orthorectification step makes it possible to correct them via a transformation of the image so that it is in the reference frame of an image that would have been acquired with an imager at a nadir angle over the whole scene and null Field of View.

Since these deformations are not simple to model and there are already orthorectified images, one solution consists of registering the image of interest on an already orthorectified one from registration points (also called GCPs for Ground Control Points). To carry out this registration, it is necessary to identify GCPs in the image to be recalibrated. In general, remarkable points are chosen, for example intersections of roofs in urban areas. Then, it is necessary to look for the equivalent points in the already orthorectified image in order to retrieve their coordinates. For example, we will identify an intersection of roofs in our image and we will try to find this intersection on Google Earth images. Thus, these points are used to calibrate a transformation model, which will then be applied to the whole image. Depending on the type of transformation, the minimum number of GCPs required may change. For example, a second-order polynomial transformation requires at least six GCPs. The quality of this registration depends on the quality of the correspondence between the pairs of points of the two images and the number of GCPs. The number of GCPs should be related to the deformations to be corrected. Indeed, the larger the deformations, the more GCPs will be required. An accuracy in the order of magnitude of the coarsest spatial resolution can be achieved between the two images.

QGIS functionality to register images:
Raster > Georeferencer > Georeferencer...

4.2.1.2.2. Radiometric preprocessing

Radiometric preprocessing consists of transforming the measurements made by the HySpex VNIR camera ("numerical count" encoded between 12 bits, "4,096 values" for each pixel and each spectral band) into reflectance factors (between 0 and 1 in the case of Lambertian surface) at the ground level for each pixel and spectral band. Reflectance factors at ground level are sought to intrinsically characterize the objects composing the scene under study by eliminating the effects of the atmosphere on the path of the measured radiation.

This step takes into account several calibration parameters associated with the sensor, the state of the atmosphere between the ground and the aircraft and the date of acquisition (the solar irradiance is not the same according to the time of day and the season). These include the absolute gain and offset calibration coefficients, the water vapor profile and the solar zenith angle [LIE 08], respectively. While the calibration coefficients are provided by the camera manufacturer and re-estimated by ONERA for each measurement campaign, the water vapor profile is either obtained from ground measurements at the same time as the acquisition or estimated by a radiative transfer model. With respect to the solar zenith angle, it is easily accessible for a given location with knowledge of the Universal Time date of the acquisition. Passing through reflectance is a delicate step because the values of most required parameters change for each spectral band of the same image. For example, the atmosphere appears blue for a human because the particles that make up the atmosphere scatter the wavelengths in the blue region more than the others such as those in the red range. This highlights the dependence of scattering on the wavelength. In the following, we will talk about spectral reflectance.

Radiometric correction is carried out with COCHISE [POU 02] (*Code de Correction Atmosphérique Hyperspectrale d'Images de Senseurs Embarqués*), a radiative transfer code developed by ONERA. The latter assumes a flat ground that implies erroneous radiometric corrections in the shadows. As a result, portions of swimming pools in the shade of buildings or trees will not be well detected. However, it can be considered that most pools are in full sun and so this assumption is valid.

4.2.2. *Reference map definition*

The developed method must be validated with a reference map. On the other hand, because we are trying to map swimming pools, we can adopt a supervised approach in which pixels extracted from the reference map can be used to define classification rules, in other words to define a training set. The reference map should contain both pools and other objects (roofs, vegetation, etc.).

4.2.2.1. *Vector layer definition*

A vector, in the sense of vector data in the context of Geographic Information Systems (GIS), contains geometric and attribute characteristics of a number of two-dimensional geometric objects of a scene from the point of view of a an observer who is usually in a satellite or aircraft. Geometric characteristics are georeferenced. ESRI's shapefile format is often used to represent vectors. There are three main types of vectors according to the geometric characteristics of the objects considered: point, polyline and polygon. The attribute characteristics can be of various types (e.g. for a pool, a real number that would be the surface of a pool or an integer that would reflect the level of negligence).

4.2.2.1.1. Choice of the entity type

The delimitation of a pool from the point of view of an observer in an airplane is like a polygon. Thus, to represent geometry, it is legitimate to consider a polygon as an entity.

4.2.2.1.2. Choice of the attributes

Pools or other objects must be identified one by one, so they must have an attribute that matches their identifier. Moreover, an object is either a pool or something else. Thus, another attribute must characterize the class of the object (pool or not-pool). Even if it is not essential, it is interesting to introduce another attribute to characterize the not-pool class, notably to specify whether the object is a roof, vegetation or a road.

> **QGIS functionality to create a vector:**
> *Layer > New > New Shapefile Layer...*

4.2.2.2. *Vector layer filling*

4.2.2.1.1. Choice of representative objects

In order to have certain diversity and thus improve the discrimination and generalization capacity of the supervised classifier, it is necessary to select different types of objects, pools but also roofs, vegetation and roads. The more representative the objects of the reference map are of those of the city, the better the method will be able to handle different cases.

4.2.2.1.2. Delineation by photointerpretation

Pools should be carefully delineated and care should be taken not to confuse them with other objects. Although they are delimited on the hyperspectral image and the spatial resolution of one meter order of magnitude is sufficient to distinguish the pools, one can use aerial images RGB (Red, Green, Blue: three broad spectral bands, analogous to what a human would see) at very high spatial resolutions (0.25-0.5 m) available on the Géoportail[3] of the IGN (*Institut National de l'information Géographique et forestière*). This can help to remove ambiguities at the edges of pools, for example with mixtures of potential materials.

> **QGIS functionality to create a polygon:**
> *- Click on the vector in the window Layers*
> *- Layer > Toggle Editing*
> *- Edit > Add Feature*

3 https://www.geoportail.gouv.fr/.

4.2.3. *Feature extraction*

4.2.3.1. *Study of the spectral reflectance*

4.2.3.1.1. Pools

A swimming pool (pool class) is an artificial basin filled with water. The spectral reflectance of a pool therefore depends on the optical properties of the water and the pool. Figure 4.2 illustrates an example of the spectral reflectance of relatively clear and very turbid water [LOI 13].

Beyond 750 nm, that is to say in the IR domain, the reflectance is almost zero. This comes from the high spectral absorption of water in this region. It is not easy to have information on the optical properties of the pool. Assuming that all pools are detected, the difference in reflectance between clear and turbid water could be exploited to identify pools that are neglected; it is reasonable to assume that a neglected pool will have more turbid water.

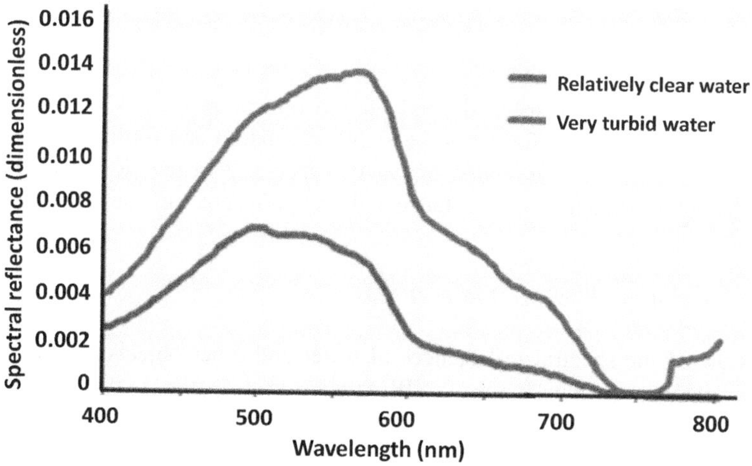

Figure 4.2. *Spectral reflectance of relatively clear and very turbid water [LOI 13]. For a color version of the figure, see www.iste.co.uk/baghdadi/qgis3.zip*

4.2.3.1.2. Other urban objects

Figure 4.3 illustrates an example of the spectral reflectance of a road, bare soil, vegetation and a roof (not-pool class) [NCR 13].

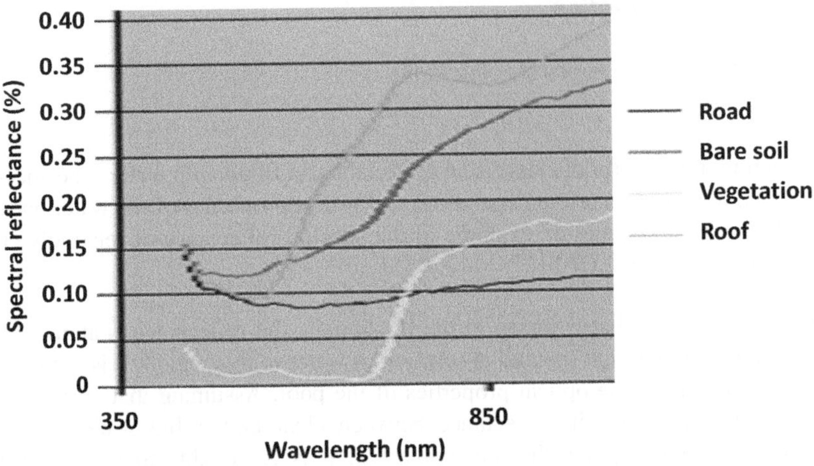

Figure 4.3. *Spectral reflectance of road, bare soil, vegetation and roof [NCR 13]. For a color version of the figure, see www.iste.co.uk/baghdadi/qgis3.zip*

These spectral reflectances depend on the optical properties of asphalt, soil components, foliage, tiles for roads, bare soil, vegetation and roofing, respectively. They are very different from that of water, especially in the IR, which *a priori* attests the interest of spectral reflectance to detect pools among other urban objects.

4.2.3.2. *Feature extraction (three cases)*

In view of the spectral reflectances of water and other objects, all the spectral bands available with the HySpex VNIR camera do not seem to be necessary. We will select spectral bands that make it possible to *a priori* differentiate pools from other objects for each pixel.

4.2.3.2.1. RGB bands

As can be seen with the RGB data at very high spatial resolution on the Geoportail, pools are already recognizable with only the visible spectrum. A first case study can therefore consider only spectral bands that are associated with typical wavelengths of red, green and blue.

4.2.3.2.2. Red, green, blue, infrared bands

Even if the pools are recognizable with the RGB data, it is not inconceivable that certain pixels are confused with other objects, especially if the water is turbid or if there are shadow phenomena. We have previously seen that in the IR, the reflectance of the water was characteristic with a value of almost zero. It is therefore interesting to add a typical IR spectral band to see its contribution.

4.2.3.2.3. Infrared band

In a third case study, only a spectral band typical of IR is considered, between 700 and 1,100 nm.

4.2.4. *Classification*

To study the potential of optical remote sensing for pool mapping in an urban environment, we consider a supervised classification approach of pixels of the reference map. It is recalled that in our case a certain number of pixels of the reference map are chosen to define the training set. The labels (pool or not-pool) of these pixels are known. A classifier is then used to define classification rules from this training set. In our case, the input attributes of the classifier are the RGB, RGB and IR or IR bands. Then, the remaining pixels (test set) are categorized to validate the method.

4.2.4.1. *Used classifiers*

Two classifiers usually used in the literature are considered: Support Vector Machine (SVM) [COR 95] and Random Forest (RF) [HO 95]. To illustrate how they work, for the sake of simple representation, simulated data presented in Figures 4.4 and 4.5 are studied. To control the illustrated cases, these data were simulated from a model that assumes that in the B and IR feature space, the pixels of the pool and not-pool classes follow Gaussian laws with non-correlated features. In particular, the pool class has low mean and variance values for the IR feature, with no particular assumption for the B feature. For the not-pool class, the mean and variance of the IR attribute are higher. These assumptions are consistent with the remarks made during the analysis of the spectral reflectance.

Each point in Figures 4.4 and 4.5 represents a pixel in the feature space of the B and IR attributes. The blue and red dots represent the pool and not-pool pixels of the reference map, respectively. They constitute a training set. The yellow and cyan dots constitute a test set. The value of the reflectance for a typical spectral band of the Blue region is on the X-axis and the value of a reflectance for a typical IR band is on the Y-axis.

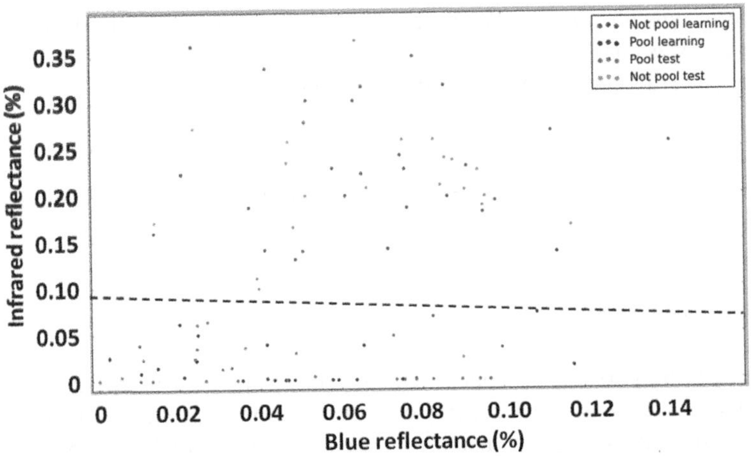

Figure 4.4. *Simulated data that illustrate how SVM works. The dashed line represents the decision boundary of a linear SVM between the data of the two classes. Below, this is the pool class; above, this is the not-pool class. For a color version of the figure, see www.iste.co.uk/baghdadi/qgis3.zip*

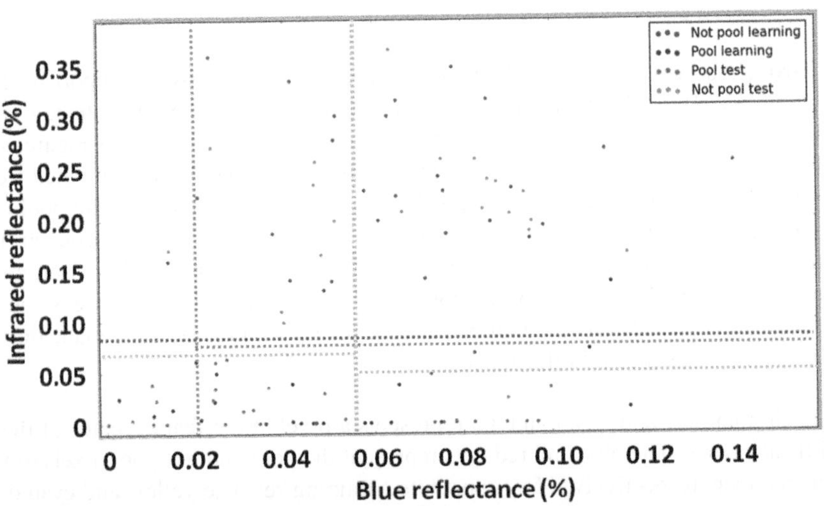

Figure 4.5. *Simulated data that illustrate how RF works. The dashed lines represent the decision boundaries of RF with at most two nodes (see decision trees in Figure 4.6). Each line represents a branch of a tree. For example, the red line should be related to the red tree (Figure 4.6). Indeed, considering this tree and in a similar way to the SVM, below it is the pool class while above it is the not-pool class. For a color version of the figure, see www.iste.co.uk/baghdadi/qgis3.zip*

4.2.4.1.1. SVM

In a two-class problem, SVM consists of finding the hyperplane that separates the training dataset from the two classes of interest. Two classes are well separated when the margin, that is to say the distance of the closest points to each of the two classes to the hyperplane, is maximal. In the case of a plane (a two-dimensional space), if the problem is linearly separable, a linear hyperplane can be found. Otherwise, the kernel trick has to be used. This consists of considering the problem in a space (of possibly greater dimension) in which it is probable that the hyperplane is linear. Formally, a nonlinear transformation is applied to the input data. This transformation is not explained but rather a kernel function calculated from this nonlinear function. The simplest example of a kernel function is the linear kernel, which reduces to the linear case without changing the space. Since it is possible that a linear separator cannot be found in the new data description space, the soft-margin technique is used that tolerates poor rankings. Consequently, a parameter C is tuned that allows us to control the compromise between the number of classification errors and the width of the margin [SCH 01].

Considering data in Figure 4.4. and a linear-kernel SVM approach, the problem is to find a line that separates red dots and blue ones. An example is shown with a black dashed line. Once the line is calculated, the points of the test set (yellow and cyan) are ranked according to their position with respect to this straight line. Thus, the points below the line belong to the pool class. It can be seen that these points are cyan, that is to say points belonging to the test sample corresponding to the pool. It is possible to use other types of kernels, for example a Gaussian one. In the latter case, the shape of the decision boundary between the data of the two classes will change.

In this case study, a Gaussian kernel is used because this classifier is commonly used in the literature for its very good performance and is easily tuned.

4.2.4.1.2. RF

The RF principle relies on decision tree learning. To overcome the known disadvantages of the decision trees (such as overfitting or sensitivity to predictor order), $\boldsymbol{B}$ decision trees are constructed using a dual sampling process: on the one hand, on the observations, a random drawing with a number of $\boldsymbol{N}$ observations equal to that of the original data (bootstrap) is applied; on the other hand, on the $\boldsymbol{p}$ predictors, only one sample $\boldsymbol{m}$ ($\boldsymbol{m} < \sqrt{\boldsymbol{p}}$ in the case of classification trees) is retained. On each sample, a decision tree is trained. The $\boldsymbol{B}$ predictions of the variable of interest are stored for each original observation. The prediction of the random forest can be achieved by a majority voting or by averaging the prediction probabilities (as implemented in the Python module Scikit-Learn [PED 11]).

In the context of the data in Figure 4.5 and RF with three trees, the problem consists of separating the set of red points and blue ones from the training set into three subsets of the same size and constructing a decision tree for each of the subsets (Figure 4.6).

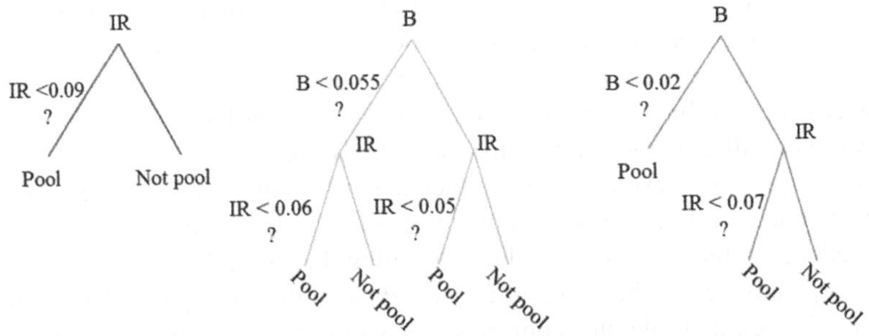

Figure 4.6. *Example of three decision trees in connection with Figure 4.5. For a color version of the figure, see www.iste.co.uk/baghdadi/qgis3.zip*

In this illustrative example, it is considered that only two bands are required for the classification, that is to say that a leaf is achieved that allows us to determine, following the decision path, whether a pool or a not-pool is found. The values of IR and B at each branch separation are calculated in such a way that the two subsets produced after the separation are relatively homogeneous from the point of view of the considered feature (B or IR herein). In general, a homogeneity criterion is optimized, for example the Gini impurity criterion [BRE 84], which is often used in the literature. Once the trees are built, each yellow and cyan point of the test set is taken and is classified with each of the trees as a function of its IR and B values. Here, the example shows that IR is almost always useful and that B is almost always uninteresting.

4.2.4.2. *Cross validation*

In the first case, 50% of the pixels of the reference map will be used for the training set.

In an operational application vision, we must ask ourselves if we will have at our disposal a number of pixels equivalent to 50% of those of the reference map in order to map the pools. In general, this is not realistic because the training set is much smaller than the test set. It is possible to select pixels of the image as is done here, but it is also possible to carry out measurements of spectral reflectance in the field in order to construct a representative library. So in a second case, 15% of the pixels will be used as a training set.

4.2.5. *Building a prediction map*

A classified pixel must have the following attributes: its reference class, the predicted class, the number of times it has been classified in the pool class, in the not-pool class and the number of times it has been tested. This information is useful for the analysis of the performance and errors of the different classifiers and the different case studies.

4.2.6. *Performance assessment*

4.2.6.1. *Global assessment*

Once the classification is completed, different statistical measures can be used to evaluate the performance of the method. The confusion matrix (Table 4.1) allows a more detailed analysis of the confusions between the classes.

		Predicted class		
		Pools	Not-pools	Total rows
Reference class	Pools	X_{11}	X_{12}	N_1
	Not-pools	X_{21}	X_{22}	N_2
	Total columns	M_1	M_2	N

Table 4.1. *Confusion matrix example*

Overall accuracy can be deduced: it is the ratio of the number of pixels well classified with respect to the total number of pixels:

$$\text{Overall accuracy (\%)} = \frac{X_{11} + X_{22}}{N} \times 100 \qquad [4.1]$$

The coefficient Kappa[4] can also be calculated: it is constructed so as to subtract the probability that the pixel is well classified by chance.

4 https://en.wikipedia.org/wiki/Cohen's_kappa.

4.2.6.2. *Spatial assessment*

From the prediction map, it is interesting to see the error areas for both classes.

QGIS functionality to see the error areas for both classes:
Layer > Create Layer > New Shapefile Layer...

From this visualization, it is possible to efficiently detect the error areas.

Error analysis is essential to have a good understanding of the behavior of the method and, in particular, problematic cases.

4.2.7. *Limits of the proposed method*

The proposed method allows us to study the potential of optical remote sensing to the cartography of swimming pools in urban areas. This is done within a reference map where some pixels compose a training sample, whereas other ones compose a test sample. Spectral signatures are built simply from observations made during spectral reflectance analysis. Classifiers are used to define classification rules on the training set and to classify the pixels of the test set. But the method has some limitations:

– we worked only with the data that are in the reference map and not the whole scene: not all the pixels of the image are classified. Even if, in the case of a classification of all pixels, it is not possible to verify exactly whether the predictions are correct, this would allow a more realistic result and the major trends for the mapping;

– in the case where all the pixels of the whole scene are classified, a map of swimming pool pixels is realized, but not of swimming pool objects. A pixel-oriented approach has been chosen for the classification, but an additional delineation step of the pools could be carried out, for example, and thus realize a real map of the swimming pools;

– the training set is composed of pixels of the image, that is to say that it is assumed that the classes of certain pixels of the image are known in advance. However, even by photointerpretation, it is not easy to access this information. In addition, performing a photointerpretation for each image may take a long time. Another solution would be to carry out spectral reflectance measurements of swimming pools and other objects in the field;

– finally, to return to one of the interests presented in the introduction in context, it is not simple to find neglected swimming pools that could be the site of development of mosquitoes that are dangerous to humans and thus to study the potential of optical remote sensing for acquiring this type of information.

4.3. Implementation of the application

This section presents a practical implementation of the detection of swimming pools on hyperspectral images acquired in the city of Toulouse (France).

4.3.1. *Software and data*

4.3.1.1. *Required software*

The raster and vector processing provided in the pool detection method uses the basic functionality of the QGIS software (version 2.18.2). This choice allows anyone to apply the processing without having to install extensions. Some steps of the method will be performed using a script in Python language requiring Numpy, OSGeo libraries (to access GDAL, OGR), Scikit-Learn (for the classification step) and Shapefile (to manipulate vector data), which can be easily installed using the conda environment (https://conda.io).

4.3.1.2. *Input data*

Data consist of a hyperspectral and atmospheric-corrected reflectance image (Figure 4.7) – **Reflectance_VNIR** – and a georeferenced (WGS 84/UTM zone 31N) Digital Surface Model (DSM) – **DSM_stereo** – centered on (375,398.91; 4,827,339.29), which extends over 1.45 km × 1.20 km with a spatial resolution of 0.125 m × 0.125 m. These data are in ENVI format.

Figure 4.7. *Illustration of the hyperspectral image used in this study acquired on a study site of* $0.496\,km \times 0.980\,km$ *on the city of Toulouse (France, X = 375,906.54; Y = 4,827,256.61). It is an atmospheric-corrected reflectance image (*$R = 632\,nm$*,* $G = 559\,nm, B = 486\,nm$*)*

4.3.2. *Step 1: creation of a georeferenced image*

<table>
<tr><th>Process</th><th>Manipulating QGIS</th></tr>
<tr><td>1. Load DSM</td><td>

In **QGIS**:

- Click on **Layer > Add Layer > Add Raster Layer**
- Load the file **DSM_stereo**

</td></tr>
<tr><td>2. Georeference hyperspectral cube using the DSM</td><td>

In the menu bar:

- Click on **Raster > Georeferencer > Georeferencer**

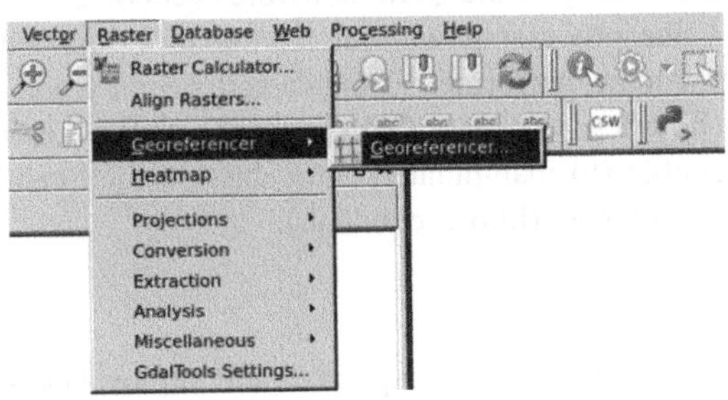

On the menu bar of the Georeferencer window:

- Click on **File > Open raster**
- Load the file **Reflectance_VNIR**

Note: If a window opens and asks for a coordinate reference system, give the same reference as the DSM. In fact, the file is not yet georeferenced.

On the menu bar of the Georeferencer window:

- Click on **Settings > Raster properties**

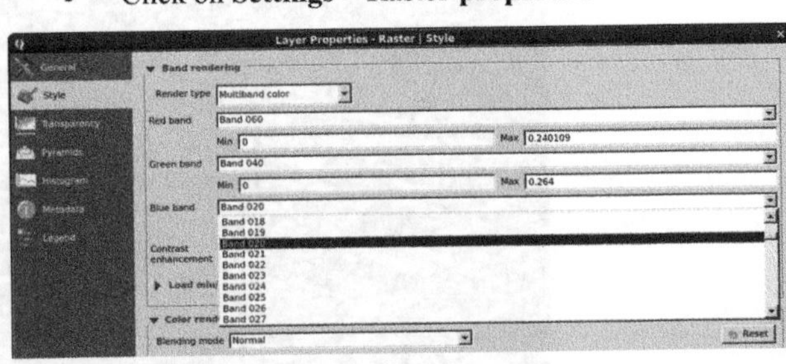

- RGB color composite can be modified. Choose band 60 for the Red band, band 40 for the Green band and band 20 for the Blue band. It provides a view similar to that which the human eye has.
- Click on **Apply** then **OK**.

</td></tr>
</table>

On the menu bar of the Georeferencer window:

- Click on **Edit > Add point**

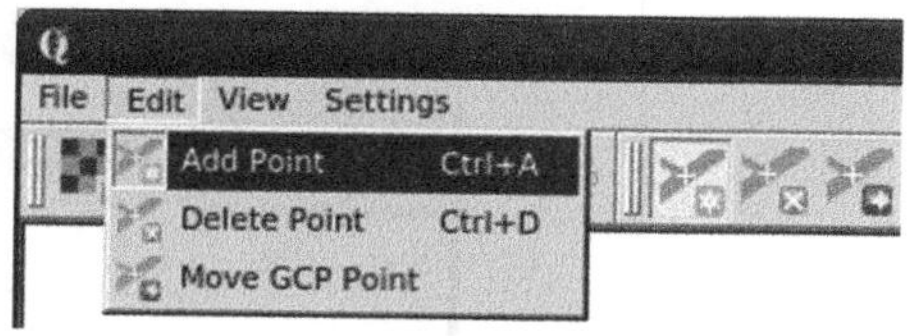

In the opened window **Enter map coordinates**:

- Click on the tab **From map canvas**

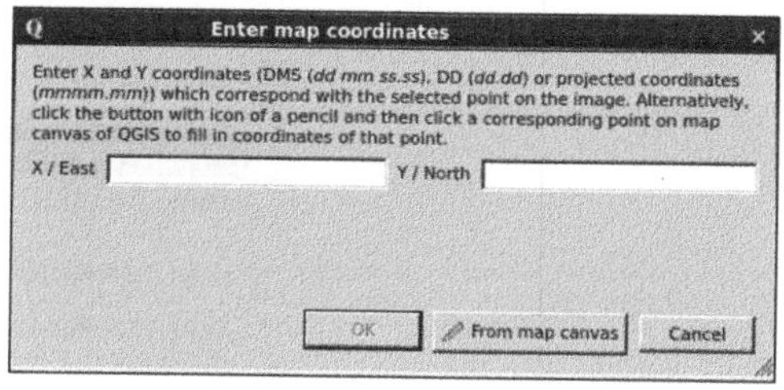

  - Click on the corresponding point on the file **DSM_stereo**.

- It may be necessary to adjust the DSM minimum values and the maximum ones to increase contrast and be more specific about the location of the DSM points.

- In the main window, right-click on **DSM_stereo** then **Properties > Style**

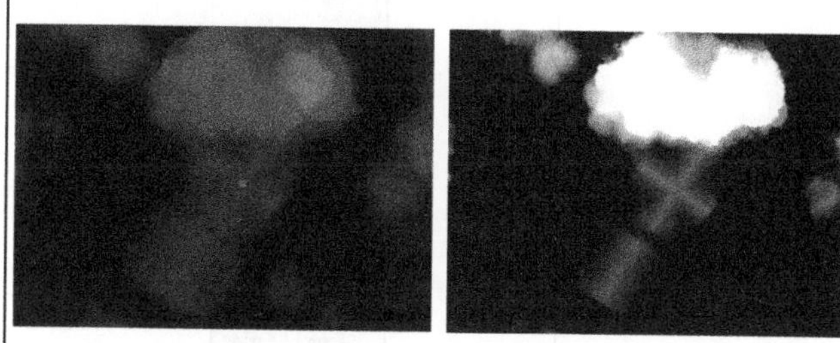

Left: example from DSM for values between 150 m and 173 m; right: example from DSM for values between 154 m and 158 m.

Note: The process of manually matching points between two data sources having different points of view and different footprints is complicated but for completeness purposes, it has been detailed here. The solution of this part is also provided with the data.

Repeat the operation in order to have a sufficient number of points.

On the Georeferencer window, on the bar menu click on:

- **Settings > Transformations Settings**

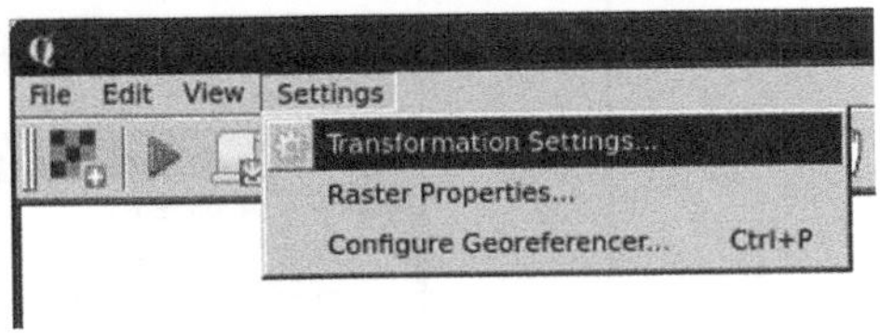

- **File > Start Georeferencing**

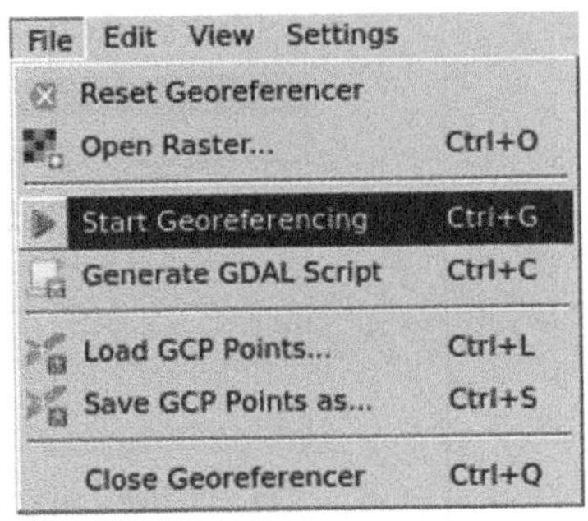

Note: In our case, the second-order polynomial that must have at least six control points is chosen.

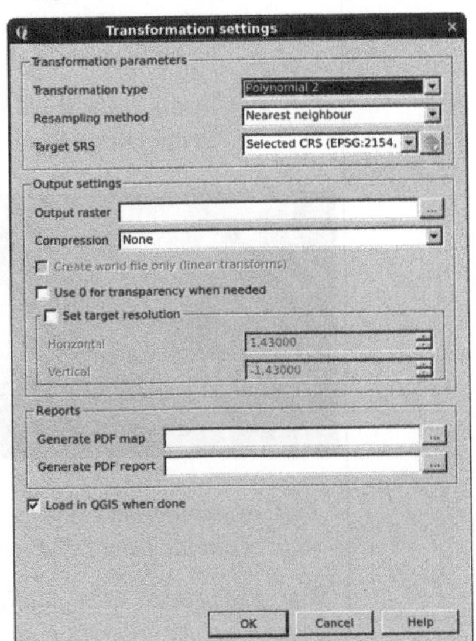

<table>
<tr>
<td></td>
<td>

Top: Example from the reflectance image; Bottom: example from the georeferenced reflectance image.</td>
</tr>
</table>

Table 4.2. *Creation of a georeferenced image (step 1). For a color version of the table, see www.iste.co.uk/baghdadi/qgis3.zip*

The georeference quality can be checked by looking at the residual error values in the GCP table (Figure 4.8). The error is the difference between the position of the point in the transformed image and the actual location specified (in the DSM). Residuals closer to zero are considered more accurate. Figure 4.9 shows the georeferenced and atmospheric-corrected hyperspectral image.

GCP table

Visible	ID	Source X	Source Y	Dest. X	Dest. Y	dX (pixels)	dY (pixels)	Residual (pixels)
☑	0	363,913	-101,262	375747	4,82696e+06	-3,14748e-08	-4,48617e-07	4,4972e-07
☑	1	904,054	-1130,03	375137	4,82756e+06	2,11283e-07	-7,81465e-07	8,09523e-07
☑	2	763,049	-609,109	375405	4,82722e+06	2,76407e-07	-6,44705e-07	7,0146e-07
☑	3	227,968	-568,627	375595	4,82731e+06	2,90885e-07	-7,9521e-07	8,46742e-07
☑	4	374,375	-967,422	375380	4,82757e+06	6,36795e-08	-6,80202e-08	9,31763e-08
☑	5	828,093	-41,335	375627	4,82682e+06	2,77127e-07	-5,40137e-07	6,07081e-07

Figure 4.8. *GCP table*

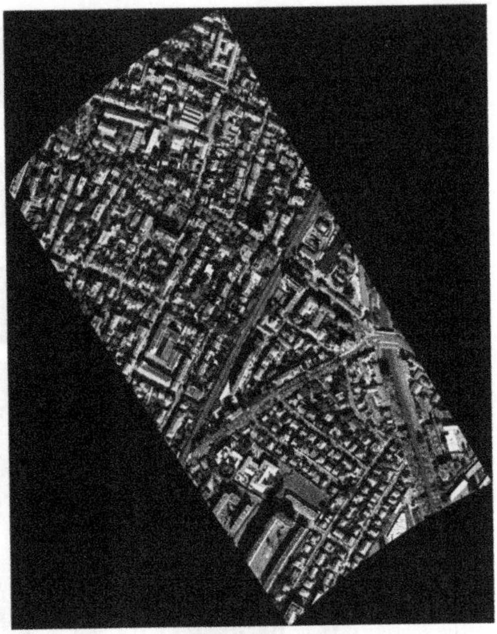

Figure 4.9. *Atmospheric-corrected and georeferenced hyperspectral image (R = 632 nm, G = 559 nm, B = 486 nm) acquired on study site of 0.496 km × 0.980 km on Toulouse (France, X = 375,906.54 ; Y= 4,827,256.61)*

4.3.3. *Step 2: building a reference map*

Processing	Manipulating QGIS
1. Build the reference vector layer	• If it is not done yet, modify RGB color composite to view the georeferenced VNIR image as above. On the menu bar: • Click on **Layer > Create layer > New Shapefile Layer…** 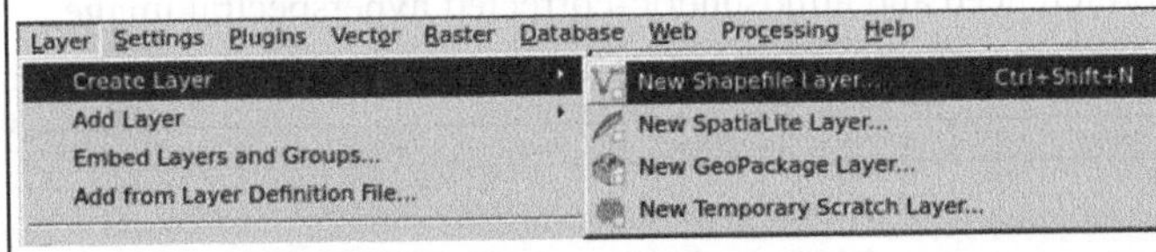 • Choose the polygon type, define the corresponding CRS (Coordinate Reference System), add the Class attribute (which can be Pools or nonPools) and the Feature one (which can be vegetation, roof and road if the class is nonPool). Both of them are Text data.

Note: Selecting training data is a simple step to achieve in practice but can dramatically affect the quality of results. Several round trips between learning and predicting are often necessary. Although each input depends on the classification problem, the classifier and the environment, two simple rules can generally be followed: (i) balance the input of the polygons of the two classes; (ii) maximize the diversification of the polygons for the two classes of interest in order to best capture the heterogeneity of the class of interest and the other landscape elements present in the non-class.

- Save the file as: **ReferenceMap**

On the Layers window, select the desired layer, then on the menu bar:

- Click on **Layer > Toggle editing**

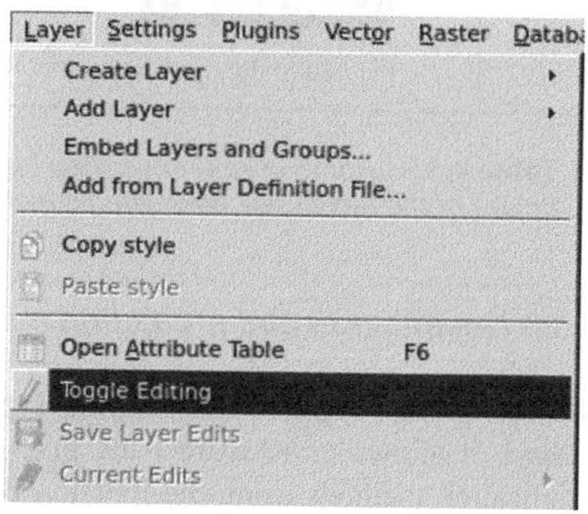

- **Edit > Add Feature**

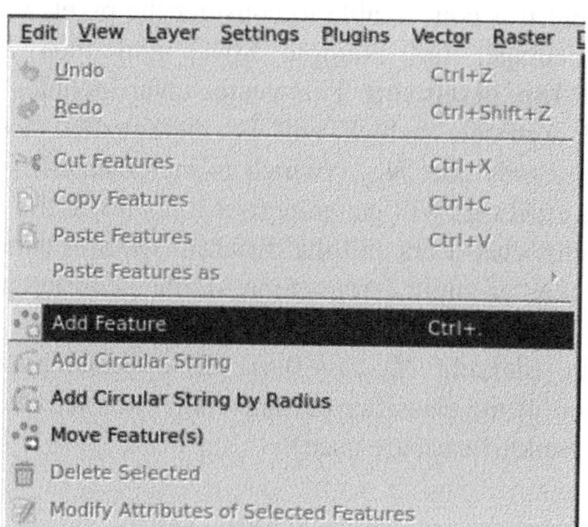

<table>
<tr>
<td></td>
<td>
Polygons are built by photointerpretation.
<ul>
<li>Define the pool or other object with the required number of points.</li>
<li>Enter the attributes. In particular, put a new value for the id attribute for each new object (for example, 1 to N if there are N objects in total). For Class and Feature attributes, it depends on whether it is a pool or another object (see the possible values defined above).</li>
<li>The more objects per class, the better (for example 30 per class).</li>
</ul>
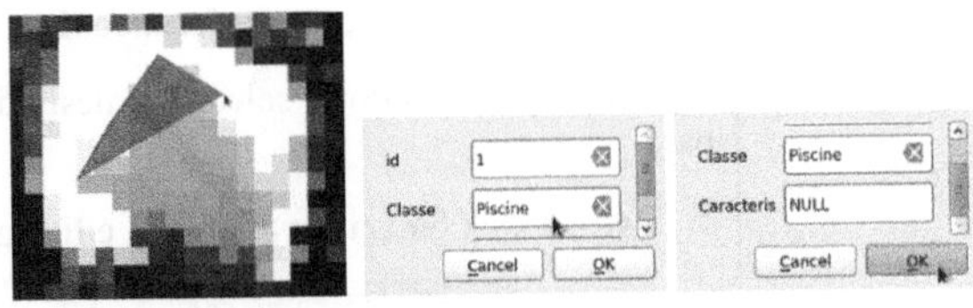

<ul>
<li>Save the modifications.</li>
</ul>
</td>
</tr>
</table>

Table 4.3. *Building of reference map (step 2). For a color version of the table, see www.iste.co.uk/baghdadi/qgis3.zip*

4.3.4. *Step 3: classification and prediction map*

The python program **PreparePoolMappingHyperspectral_VNIR.py** is used. On the one hand, it allows us to extract the pixels that will be required for the different classification methods using the function **BuildReferenceMapPixel** (only pixels that are entirely inside a polygon given by the reference map (Figures 4.10 and 1.11) but it is conceivable to consider the pixels intersecting the polygon with a given percentage, for example 80%) and thus a vector layer is realized: **ReferenceMap_pixels.shp**. This vector layer includes the following new attributes: Pools and nonPools (which will be incremented according to the classification result) and a new one: N_{cur} (which is incremented if the pixel belongs to the test set). These attributes will generate useful results in the interpretation of the precision of the various classifiers and the different spectral bands (confusion matrixes). This program takes as input parameters: **-I** the georeferenced image, **-S** the reference map, -N the number of Monte Carlo iterations (default: $N_{iter} = 30$), **-trS** the training sample size (default: $N_{test} = 0.5$). An optional parameter is also available: **-O** optimization of the classifier parameters (if not activated, the parameters defined by default by Scikit-Learn are used).

Figure 4.10. *Overlapping of the reference map and the georeferenced hyperspectral image: the swimming pools are blue and the non-pools are red. For a color version of the table, see www.iste.co.uk/baghdadi/qgis3.zip*

Figure 4.11. *Illustration of selected pixels within polygons. For a color version of the table, see www.iste.co.uk/baghdadi/qgis3.zip*

The program is launched in a terminal: python PreparePoolMapping Hyperspectral_VNIR.py –I georeferenced_image –S reference Map.shp.

NOTE.– For now, the program only supports ENVI and GTiff formats for the georeferenced images.

From the vector layer **ReferenceMap_pixels.shp**, a prediction map is realized by classifier and by spectral band(s) (**ReferenceMaps_pixel_Classifier_ Bands.shp**). At each iteration, the columns "Pools", "nonPools" and "N_{cur}" of this vector layer are updated, in order to have the necessary data for the confusion matrix. The confusion matrix files averaged by the iteration number (**Bands_Classifieur_ResultsConfusionMatrixRepresentative.txt**) and the global precision and the kappa file of each the classifiers for each spectral domain (**ResultsRepresentative.txt**) are generated.

<table>
<tr><th>Process</th><th>Manipulating QGIS</th></tr>
<tr><td>1. View prediction map using a color gradient</td><td>
In the Layers window:
<ul><li>Right-click on the appropriate layer</li><li>Click on Properties</li></ul>
On the Layer Properties window:
<ul><li>Click on the tab Style
<ul><li>Select Graduated</li></ul></li></ul>
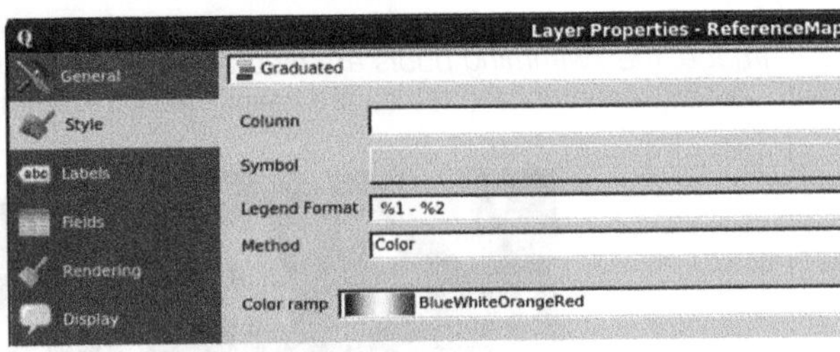

<ul><li>Write the following formula for the new column clicking on the button ε :</li></ul>
CASE WHEN ("Class" IS 'Pool') THEN (0.5 + 0.5 * toreal("Pools")/toreal("N_cur")) ELSE (0.5 – 0.5 * toreal("nonPools")/toreal("N_cur")) END

This formula allows you to store the classification results of the two classes on a common scale: from 0 to 0.5 the nonPools class classification results are given; from 0.5 to 1 the Pool ones are given; the closer to 0.5, the worse the classification.

Choose a color ramp like RdBu (which ranges from red to blue: red is associated with the nonPool class, whereas blue is associated with the Pool one) that allows you to get a dark color if the class is well found, a near-white color otherwise.
</td></tr>
</table>

Table 4.4. *Classification and prediction map (step 3).*
For a color version of the table, see www.iste.co.uk/baghdadi/qgis3.zip

Figure 4.12 shows that, regardless of the chosen spectral domain, RF is more robust than SVM with a Gaussian kernel. Indeed, on this illustrative example, the tile is always well classified by the RF, whether it is shaded or sunny (the pixels are dark red). On the other hand, considering the sunny tile, the worst results are given by SVM for RGB or RGB-IR: the lighter the red pixels are, the worse the classification. The classifier error is thus displayed for a given group of spectral bands: this classifier tends to classify these sunlit tile pixels as pool pixels, while the classification of the sunlit tile pixels remains good for the IR band. However, this result is to be qualified: Figure 4.13 shows that for the IR band, regardless of the classifier, the pixels on the pool boundary are less well classified (they are light blue). The misclassification of these pixels can be explained by the fact that they contain the spectral signature of mixed materials. Nevertheless, considering the object instead of the pixels, the swimming pool is correctly classified.

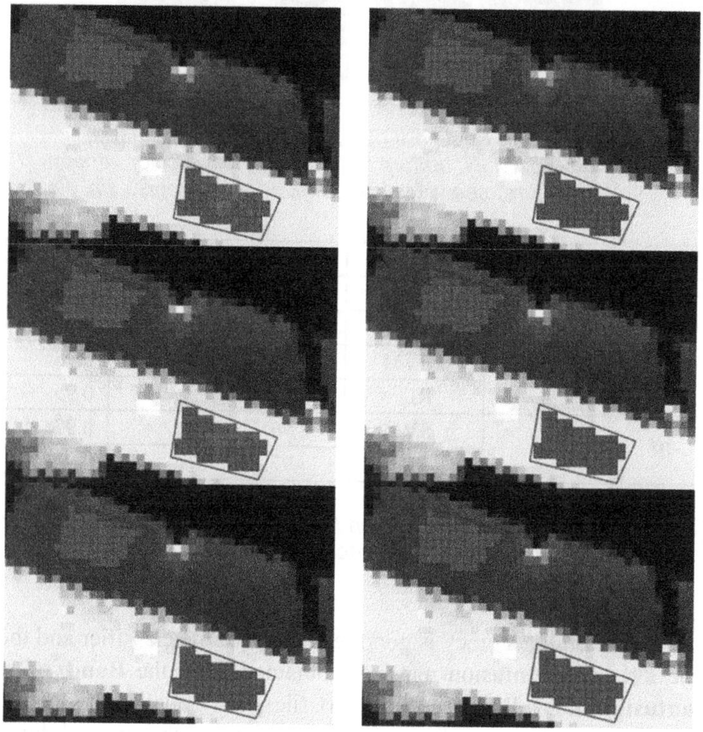

Figure 4.12. *Not-pools are red. Top left: SVM RGB; top right: RF RGB; middle left: SVM RGB-IR; middle-right: RF RGB-IR; bottom left: SVM IR; bottom right: RF IR. For a color version of the figure, see www.iste.co.uk/baghdadi/qgis3.zip*

Considering all the pixels of the reference map, Table 4.5 (produced from the **ResultsRepresentative.txt** file) shows us that SVM gives better overall accuracy than RF in the RGB and RGB-IR domains for this particular training set. The reader might get slightly different conclusions without fundamentally changing the essence of this chapter. It may be noted that the differences between the two techniques are minimal, as noted by many authors in the literature.

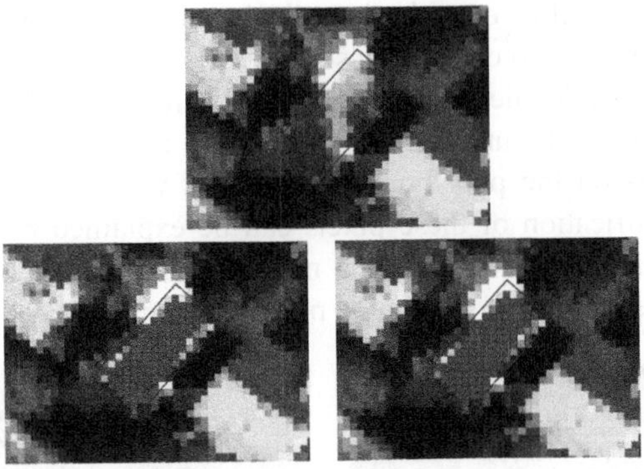

Figure 4.13. *Pools are blue. Top: reflectance image; bottom left: SVM IR; bottom right: RF IR. For a color version of the figure, see www.iste.co.uk/baghdadi/qgis3.zip*

Spectral band(s)	Classifier	Overall accuracy (%)	Kappa
RGB	SVM	99.2	0.98
	RF	96.9	0.94
RGB-IR	SVM	98.9	0.98
	RF	97.6	0.95
IR	SVM	96.7	0.93
	RF	97.1	0.94

Table 4.5. *Overall accuracy (%) and Kappa for each of the classifiers and each of the spectral bands studied*

Although the overall accuracy is good regardless of the classifier and the spectral domain, the averaged confusion matrices (obtained from the **Bands_Classifieur_ResultsConfusionMatrixRepresentative.txt** files) may qualify this result. Indeed, Tables 4.6–4.11 show that regardless of the classifier and the spectral domain, there is a tendency to overestimate the number of pixels of the pool class.

		Predicted class	
		Pools	Not-pools
Reference class	Pools	579.17	0.87
	Not-pools	10.20	783.54

Table 4.6. *Confusion matrix averaged by the number of iterations of SVM applied to RGB bands*

Reference class		Predicted class	
		Pools	Not-pools
	Pools	575.23	4.80
	Not-pools	37.87	755.87

Table 4.7. *Confusion matrix averaged by the number of iterations of RF applied to RGB bands*

Reference class		Predicted class	
		Pools	Not-pools
	Pools	578.57	1.47
	Not-pools	13.40	780.33

Table 4.8. *Confusion matrix averaged by the number of iterations of SVM applied to RGB-IR bands*

Reference class		Predicted class	
		Pools	Not-pools
	Pools	575.5	4.53
	Not-pools	28.87	764.87

Table 4.9. *Confusion matrix averaged by the number of iterations of RF applied to RGB-IR bands*

Reference class		Predicted class	
		Pools	Not-pools
	Pools	567.20	12.84
	Not-pools	23.43	761.30

Table 4.10. *Confusion matrix averaged by the number of iterations of SVM applied to IR band*

Reference class		Predicted class	
		Pools	Not-pools
	Pools	567.40	12.63
	Not-pools	27.47	766.27

Table 4.11. *Confusion matrix averaged by the number of iterations of the RF applied to IR band*

4.4. Bibliography

[AKB 16] AKBARI D., HOMAYOUNI S., SAFARI A. *et al.*, "Mapping urban land cover based on spatial-spectral classification of hyperspectral remote-sensing data", *International Journal of Remote Sensing*, vol. 37, no. 2, pp. 440–454, 2016.

[BRE 84] BREIMAN L., FRIEDMAN J., STONE C.J., *et al.*, *Classification and Regression Trees*, CRC Press, Boca Raton, 1984.

[COR 95] CORTES C., VAPNIK V., "Support-vector networks", *Machine Learning*, vol. 20, no. 3, pp. 273–297, 1995.

[GAL 09] GALINDO C., MORENO P., GONZÁLEZ J., *et al.*, "Swimming pools localization in colour high-resolution satellite images", *2009 IEEE International Geoscience and Remote Sensing Symposium (IGARSS 2009)*, vol. 4, Cape Town, South Africa, 2009.

[HO 95] HO T.K., "Random decision forests", *Proceedings of the Third International Conference on Document Analysis and Recognition*, vol. 1, pp. 278–282, 1995.

[KIM 11] KIM M., HOLT J.B., EISEN R.J., *et al.*, "Detection of swimming pools by geographic object-based image analysis to support West Nile Virus control efforts", *Photogrammetric Engineering & Remote Sensing*, vol. 77, pp. 1169–1179, 2011.

[LE 10] LE MONITEUR, "Piscines privées: la France, 2e marché au monde derrière les Etats-Unis", *Le Moniteur*, available at http://www.lemoniteur.fr/article/ piscines-privees-la-france-2e-marche-au-monde-derriere-les-etats-unis-3178576, 2010.

[LIE 08] LIER P., VALORGE C., BRIOTTET X., *Imagerie spatiale: des principes d'acquisition au traitement des images optiques pour l'observation de la terre*, Cepadues, Toulouse, 2008.

[LOI 13] LOISEL H., VANTREPOTTE V., JAMET C., *et al.*, "Challenges and new advances in ocean color remote sensing of coastal waters", in ZAMBIANCHI E. (ed.), *Topics in Oceanography*, InTech, London, 2013.

[NCR 17] NCRST, Road centerlines from hyperspectral data, Technical Report, available at http://ncrst.digitalgeographic.com/resources/easyread/Hyper Center-lines/first.html, 2017.

[PED 11] PEDREGOSA F., VAROQUAUX G., GRAMFORT A., *et al.*, "Scikit-Learn: machine learning in python", *Journal of Machine Learning Research*, vol. 12, pp. 2825–2830, 2011.

[PHI 17] PHILLIPSON A., "Ils cachent l'existence de leurs piscines au fisc... et sont démaqués par Google Maps", *La Dépêche*, available at http://www.ladepeche.fr/article/2017/03/06/2530274-cachent-existence-piscines-fisc-sont-demasques-google-maps.html, 2017.

[POU 02] POUTIER L., MIESCH C., LENOT X., *et al.*, COMANCHE and COCHISE: two reciprocal atmospheric codes for hyperspectral remote sensing, *2002 AVIRIS Earth Science and Applications Workshop Proceedings*, Pasadena, California, USA, 2002.

[ROD 14] RODRIGUEZ-CUENCA B., ALONSO M.C., "Semi-automatic detection of swimming pools from aerial high-resolution images and LiDAR data", *Remote Sensing*, vol. 6, pp. 2628–2646, 2014.

[SCH 01] SCHOLKOPF B., SMOLA A.J., *Learning with Kernels: Support Vector Machines, Regularization, Optimization, and Beyond*, MIT Press, Cambridge, 2001.

5

Automation of Workflows for the Installation of a Wind Farm

5.1. Automation of workflows

The implementation of automation of workflows now allows new forms of interactions between users and spatial data, both for analysis and interoperability. Software solutions like ESRI *Model Builder* or *Extract Transform Load* enable rapid implementation of concatenation models of sequences of geoprocessing tools, which are particularly useful and often unavoidable at a time of data proliferation.

The objective of an automated workflow is to apply a series of varied processings and/or transformations (attribute, geometric, projection, conversion, etc.) onto a set of data by following a defined path or by iteration. Consolidated in a single model, the processes are easier to run, take less time and can be reused on other datasets.

QGIS proposes (since version 2.2) an intuitive graphical modeling interface for the implementation of automated workflows. Accessible via the graphical interface of the SEXTANTE library, the graphical modeler makes it possible to create complex models using a simple and easy-to-use interface. The other advantage of the graphic modeler is the ability to combine different libraries (GDAL, GRASS, SEXTANTE, Saga, etc.). It should be noted, however, that the QGIS graphic modeler is not fully stabilized and that the implementation of robust and operational models requires rigor and patience.

Chapter written by Boris MERICSKAY.

5.2. Automation of workflows for the installation of a wind farm in Brittany

The objective of this chapter is to set up, via the QGIS graphical modeler, an automated workflow based on vectorial datasets to identify the optimal areas for the installation of a new wind farm in Brittany. The use of this tool makes it possible to mobilize different algorithms from the SEXTANTE and GDAL libraries.

The analysis model, which aims to identify the optimal wind zones, is based on a set of attributes and spatial criteria relating to the organization of the area, the regulations in force and the wind potential; summarized in Table 5.1.

	Type	Processing	Dataset
Population	Exclusion	Symmetric difference	Population grid
Protected areas	Exclusion	Union and difference	Regional natural park
Wind farm	Exclusion	Buffer and difference	Wind farm
Regional wind	Inclusion	Intersection	Wind energy regional plan
Wind energy	Inclusion	Intersection	Wind power density
Power line	Inclusion	Buffer and intersection	Power line

Table 5.1. *Summary of model criteria*

In addition to the principles and functionalities of concatenation and automation processings, the chapter also discusses the use of Geographic Web Services to download spatial data directly into QGIS, which is becoming more widespread within GIS environments (Web Feature Service (WFS)[1] and QuickOSM extension[2]).

Figure 5.1 illustrates the succession of processings required for the identification of optimal areas for the installation of a new wind farm. To facilitate its reading, the processings are grouped into six main steps, from the download of the data to visual verification:

– preparation of population grid dataset;

1 A Web Feature Service (WFS) is an Open Geospatial Consortium standard which makes it possible to directly download vector datasets into GIS software. Unlike Web Map Service (WMS) which only allows for visualization of data *via* streams, WFS retrieves datasets to allow the user to change formatting or to perform transformations and analyses.

2 QuickOSM is a QGIS plugin which makes it possible to directly download data for use in QGIS from the OpenStreetMap database via Overpass API.

– identification of inhabited areas;

– consideration of protected areas;

– consideration of the regional wind energy policy;

– consideration of the proximity of power lines;

– visual verification.

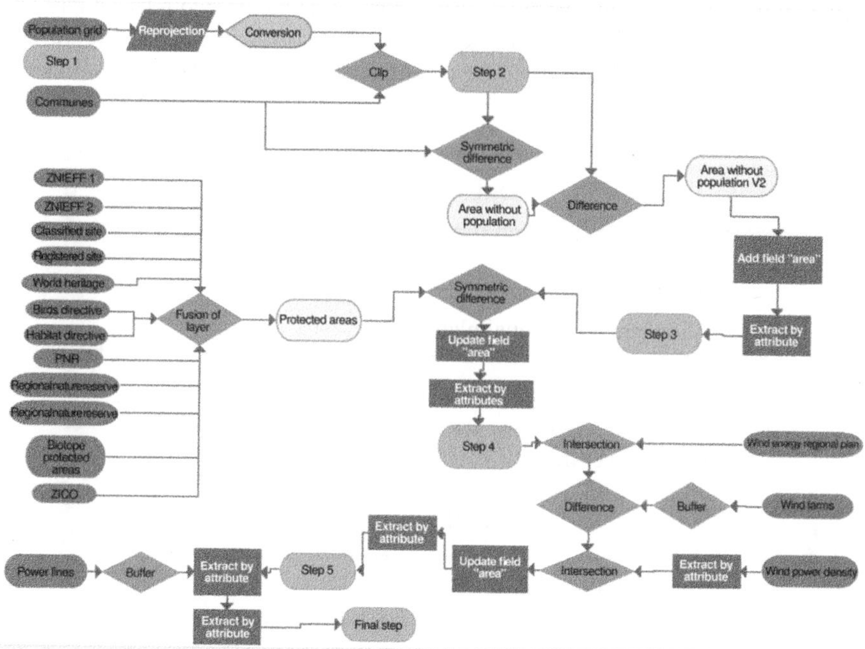

Figure 5.1. *Global workflow of chapter*

5.2.1. *Step 1: download data with WFSs*

The first step is to download the different datasets needed to carry out the implementation study. The 18 datasets are produced by several organizations (IGN[3], INSEE[4], Brittany region, Brittany DREAL[5], Minister of the Environment and

---

3 National Institute for Geographic and Forestry Information (*Institut Nationale de l'Information Géographique et Forestière*).

4 National census bureau (*Institut National de la Statistique et des Etudes Economiques*).

5 Regional organization of land use planning (*Direction Régionale de l'Environnement, de l'Aménagement et du Logement*).

OpenStreetMap) and can be downloaded by using several methods (direct download, WFS and QuickOSM extension).

5.2.1.1. *Using the WFSs of GeoBretagne*

GeoBretagne is the institutional spatial data infrastructure (SDI) of the Brittany region. This SDI centralizes more than 1,500 open and interoperable spatial datasets. It offers a series of Geographic Web services (search, visualization, cataloging and downloading). The idea is to mobilize WFS as much as possible to retrieve datasets directly from QGIS for the implementation of automated chains. Figure 5.2 summarizes the different datasets to be downloaded by this method as well as the associated WFS services.

DREAL WFS

- Municipalities
- Wind farms
- Wind energy regional plan

Brittany region WFS

- Wind power density

CARMEN WFS

- Regional natural park
- ZICO
- ZNIEFF 1
- ZNIEFF 2
- Biotope areas
- Regional nature reserve
- Regional nature reserve
- World heritage
- Registered sites
- Classified sites
- Habitat directive
- Birds directive

Figure 5.2. *WFS and associated datasets*

When downloading datasets via WFS, you should project the layers using the coordinate reference system (CRS) Lambert 93 – EPSG 2154. Once the datasets are loaded into QGIS, even if it is possible to mobilize the downloaded layers of the WFS (stored in the cache) within the graphic modeler, it is also advisable to save the layers locally to avoid possible errors and conflicts in the execution of the processing chains.

5.2.1.2. *Using the QuickOSM plugin to download OpenStreetMap data into QGIS*

QuickOSM is a QGIS plugin, available in the QGIS plugins repository, which offers the possibility to download and use OpenStreetMap data in QGIS. The main advantage of this extension is the intuitive dialog boxes that allow you to quickly build customized queries at the attribute (e.g. downloading certain types of roads, power lines, bus stops or administrative boundaries) and spatial level (right-of-way, layer, places, etc.).

OSM data are semantically described by a multitude of tags based on attributes comprising a key and a value. For example, a highway is qualified by the generic key *highway* and its value is *motorway*, a main road is qualified by the *Highway* key and its value is *Primary*. This system of semantic qualification via keys allows for permanent scalability and, in particular, multiple, advanced qualifications (e.g. a road can be characterized by its type, direction, the number of lanes, maximum speed, etc.).

In the context of this chapter, the idea is to be able to download the power lines of Brittany. To do this, it is necessary to specify: (1) the key – *Power* (refers to everything that is energy related); (2) the value – *Line* (refers to the power lines); (3) and to define the spatial extent of the download (*Municipalities layer*). By default, the downloaded layers use WGS84 as the CRS (EPSG 4326). In the context of this analysis (in France), it is necessary to use Lambert 93 (ESPG 2154) as the common CRS (it is therefore necessary to reproject the layer before integrating it into the model).

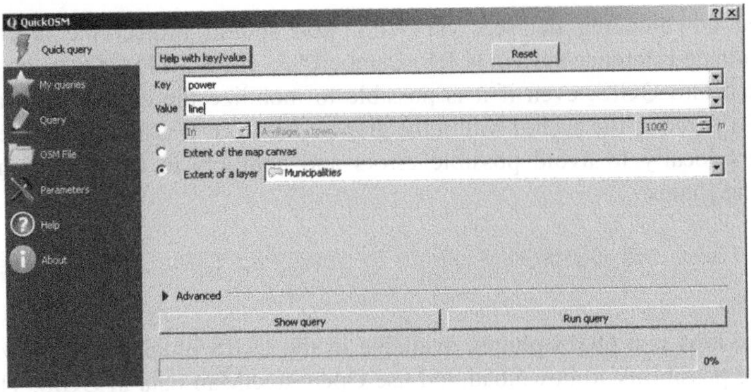

Figure 5.3. *Query for download Power lines of Brittany*

5.2.2. *Step 2: preparation of the population grid dataset*

The population grid dataset available on the INSEE website cannot be directly integrated into complex models (MIF-MapInfo format, CRS not defined, national scale). The first workflow of this chapter (Figure 5.4) therefore aims to normalize this input dataset (format, CRS, extent). This operation can be compared to a typical extraction/transformation operation of ETL software. It consists of three steps: (1) reprojection in Lambert 93; (2) conversion into a shapefile format; (3) geometric clip for isolated population grid of Brittany.

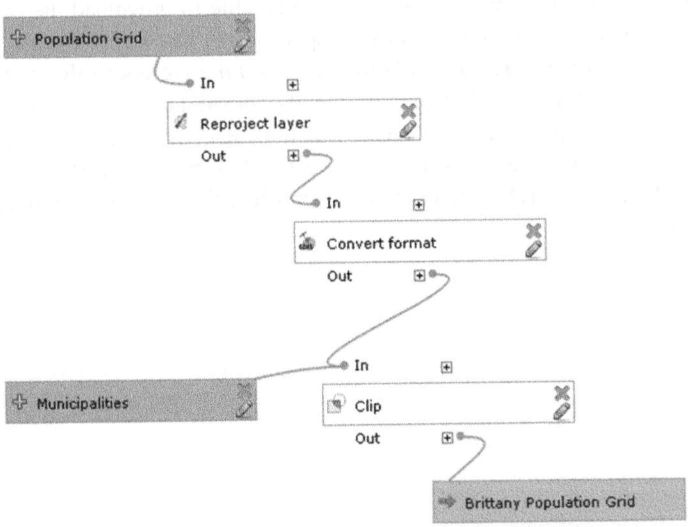

Figure 5.4. *Workflow for preparation of grid population dataset*

QGIS functionalities:
- Project layer: *QGIS Geoalgorithms > Vector general tools > Reproject layer*
- Dataset format conversion: *Library GDAL/OGR >Conversion > Convert*
- Clip: *QGIS Geoalgorithms > Vector overlay tools > Clip*

5.2.3. *Step 3: identification of inhabited areas*

Once the population grid datasets have been prepared, the third step is to begin the analysis by adding a first criterion in the model (exclusion), the presence of population. The legislation concerning the installation of wind turbines stipulates that they cannot be built within 500 m of a building (house, farm, agricultural shed, etc.). In order to satisfy this first criterion, the INSEE data of 1 km grids makes it possible to visually identify the void areas (zones without tiles) as illustrated in Figure 5.5. Even if the INSEE data are to be used with caution (statistical methods), they nevertheless offer a way of modeling the population of interest at the regional scale.

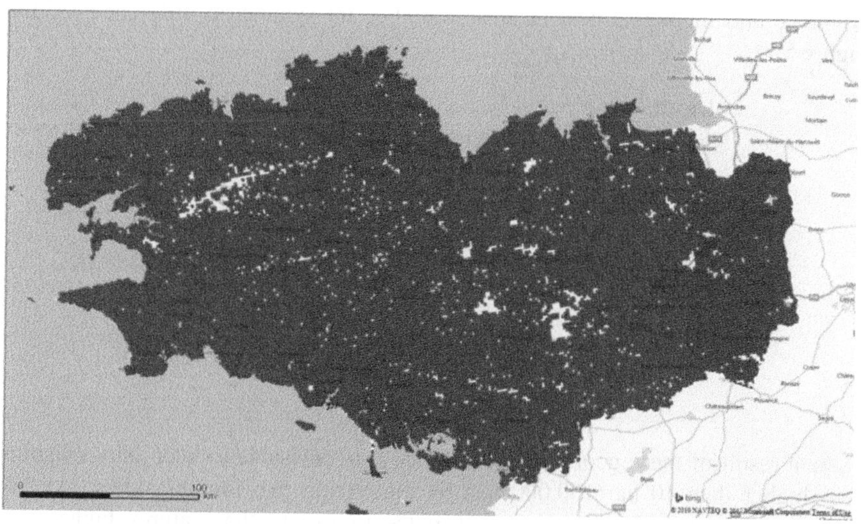

Figure 5.5. *Brittany population grid. For a color version of the figure, see www.iste.co.uk/baghdadi/qgis3.zip*

To vectorize inhabited areas, it is necessary to mobilize an adapted vector overlay tool: the **symmetrical difference**[6] between the municipalities layer (Input layer) and the population grid layer (Difference layer).

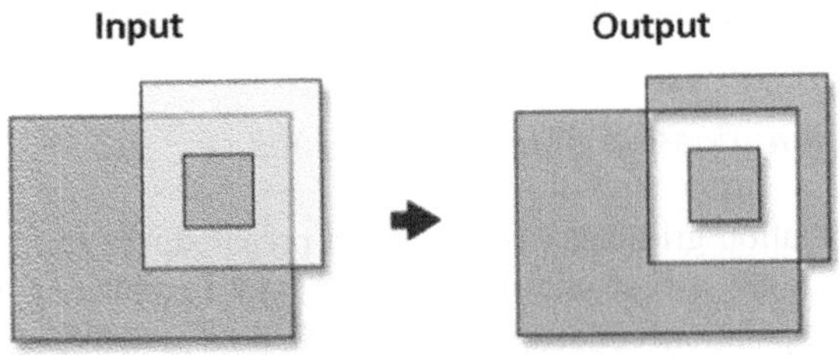

Figure 5.6. *Principle of symmetrical difference. For a color version of the figure, see www.iste.co.uk/baghdadi/qgis3.zip*

Depending on the topological coherence of the datasets, it may be relevant, following symmetric differentiation, to apply **difference**[7] (between symmetrical difference and municipalities) to erase the aberrant symmetric difference residues (Figure 5.7).

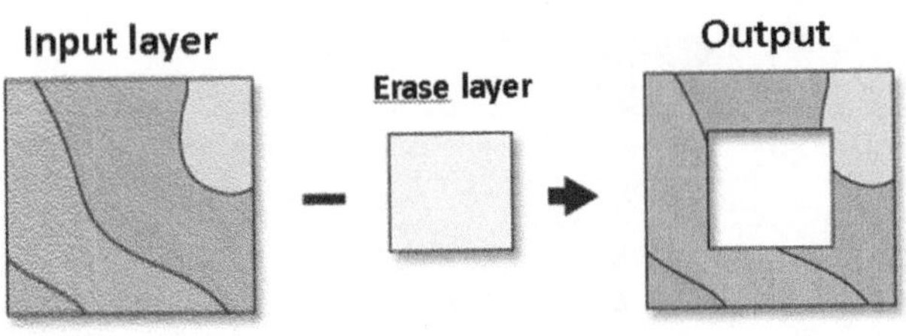

Figure 5.7. *Principle of difference. For a color version of the figure, see www.iste.co.uk/baghdadi/qgis3.zip*

As a result of these geometric operations, it is advisable to keep only inhabited areas higher than 10 ha (100,000 m^2) by mobilizing two functionalities: (1) Add geometry columns; (2) extract by attributes to isolate the features higher than 100,000 m^2 (Figures 5.8 and 5.9).

---

6 This algorithm creates a layer containing features from both the Input and Difference layers but with the overlapping areas between the two layers removed.

7 This algorithm extracts features from the Input layer that fall outside, or partially overlap, features in the Difference layer.

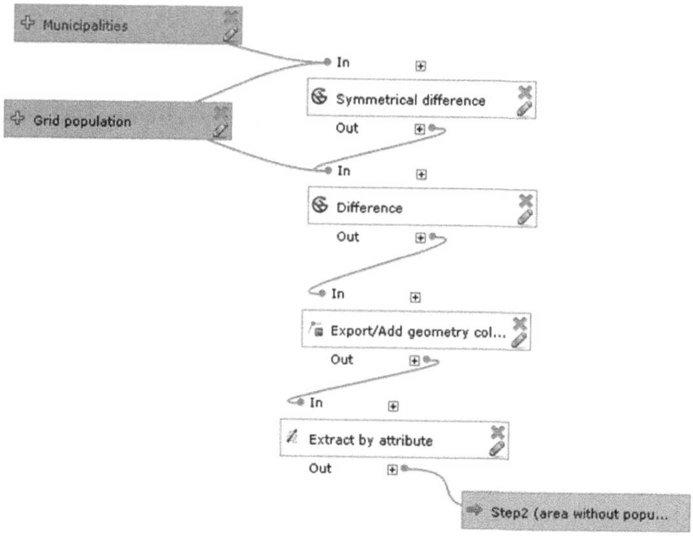

Figure 5.8. *Workflow for identification of inhabited areas*

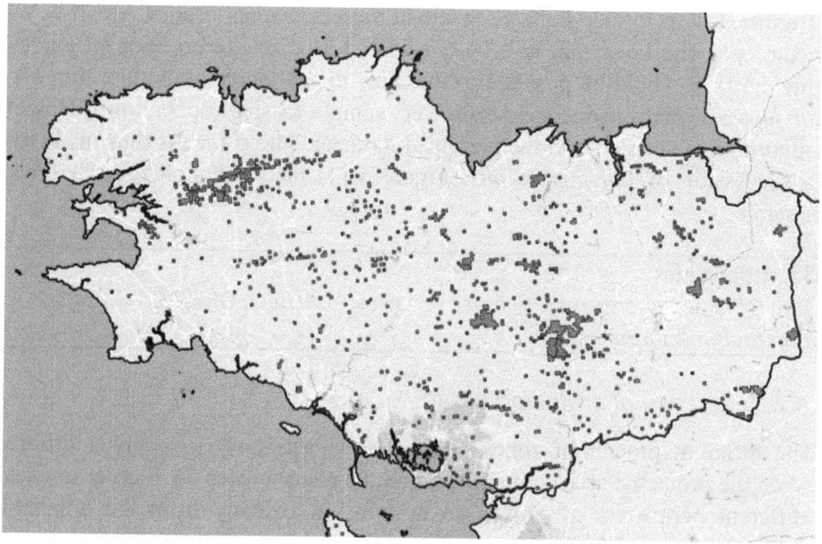

Figure 5.9. *Mapping result of uninhabited areas. For a color version of the figure, see www.iste.co.uk/baghdadi/qgis3.zip*

QGIS functionalities:
- Symmetrical difference: *QGIS geoalgorithms > Vector overlay tools > Symmetrical difference* **OR** *SAGA > Vector polygon tools > Symmetrical difference*
- *Difference: QGIS geoalgorithms > Vector overlay tools > Difference* **OR** *SAGA > Vector polygon tools > Difference*
- Area calculation: *QGIS geoalgorithms > Vector table tools > Export / Add geometry columns*
- Select features by area: *QGIS geoalgorithms > Vector selection tools > Extract by attribute*

5.2.4. *Step 4: consideration of protected areas*

Once the zones are isolated, the third model considers protected areas (as exclusion criterion) for the establishment of a new wind farm. The legislation prohibits the installation of wind turbines within various protection areas (ZNIEFF, regional natural parks, nature reserves, etc.).

5.2.4.1. *Union of different protected areas*

The first task is to merge the 12 layers of protected areas (from CARMEN WFS) to create only one layer that will be considered as a spatial criterion of exclusion (Figure 5.10). It would be possible to mobilize the geoprocessing algorithm *Merge vector layers*, which, in theory, seems very simple (a single operation). However, this algorithm is not stable in the graphical modeler. The choice is thus made to opt for a succession of *Polygon unions* (Figure 5.11) that, although more tedious, is more stable.

QGIS functionalities:
- Union: *QGIS geoalgorithms > Vector overlay tools > Union* **OR** *SAGA > Vector polygon tools > Polygon union*

5.2.4.2. *Difference of protected areas*

The different protection zones merged, it is necessary to apply a difference (between the protected areas union layer and the result of step 2) in order to exclude the different perimeters of protection in force in Brittany from the uninhabited zones. In addition and in order to eliminate the residuals of the difference between the two layers, it is advisable to update the surface field and then make an extraction by attribute by applying a selection based on a minimum surface of 100,000 m^2 (Figures 5.12 and 5.13).

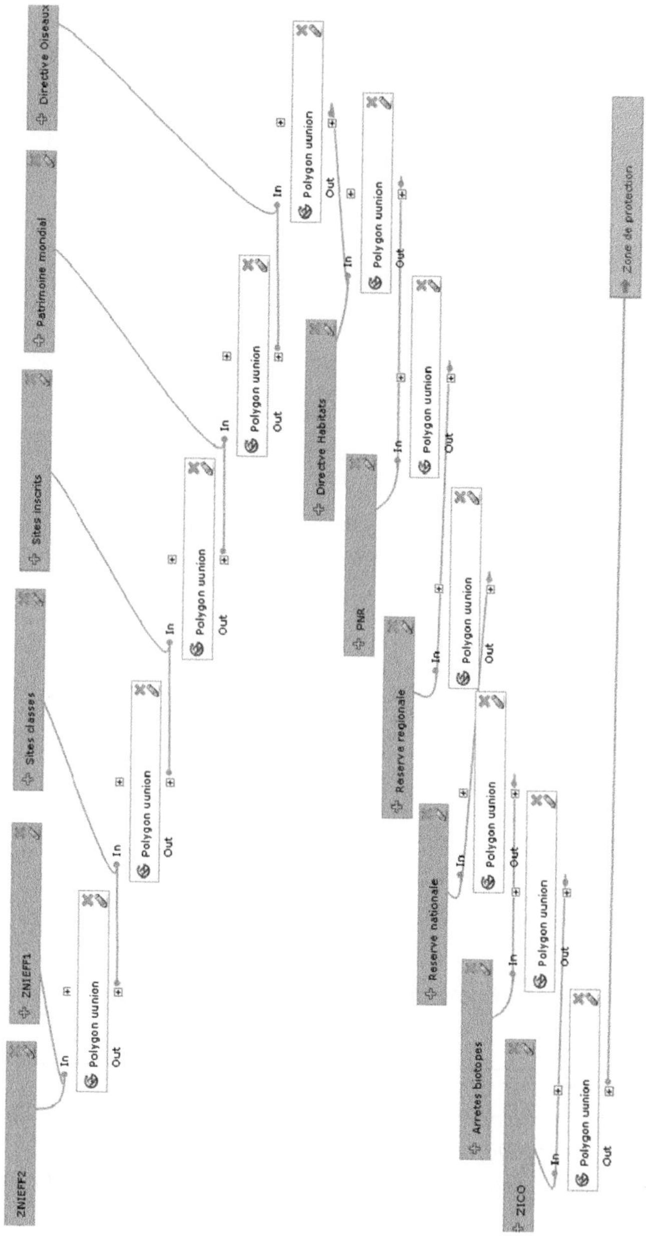

Figure 5.10. *Workflow of protected areas union*

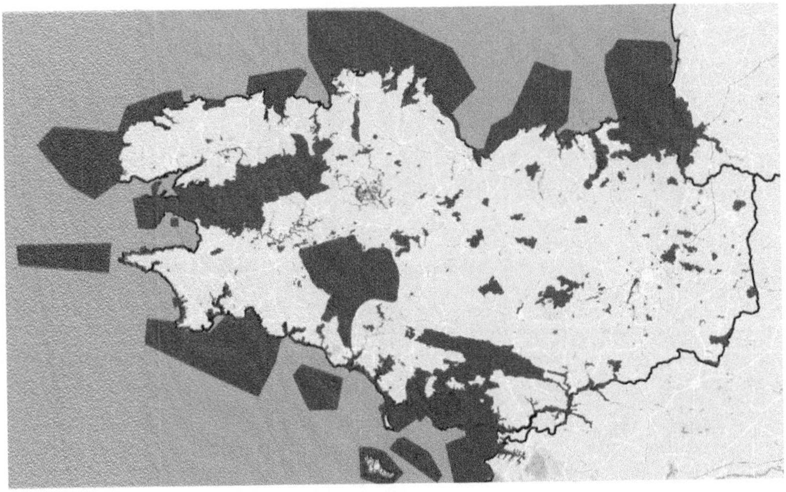

Figure 5.11. *Mapping result of different protected areas union. For a color version of the figure, see www.iste.co.uk/baghdadi/qgis3.zip*

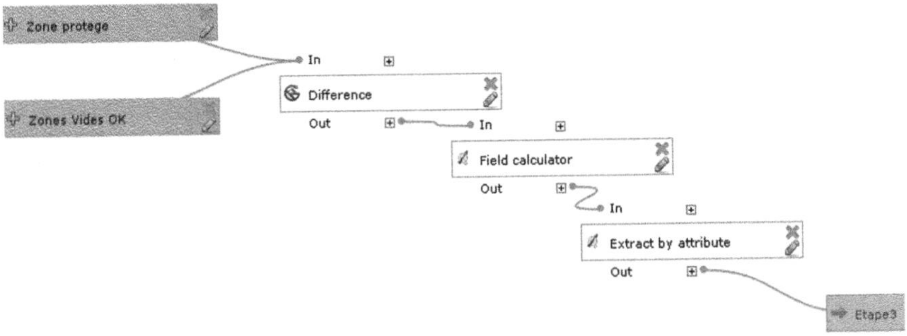

Figure 5.12. *Workflow of the difference of protected areas*

QGIS functionalities:

- Difference: *QGIS geoalgorithms > Vector overlay tools > Difference* **OR** *SAGA > vector polygon Tools > Difference*
- Area calculation: *QGIS geoalgorithms > Vector table tools > Export / Add geometry columns*
- Select features by area: *QGIS geoalgorithms > Vector selection tools > Extract by attribute*

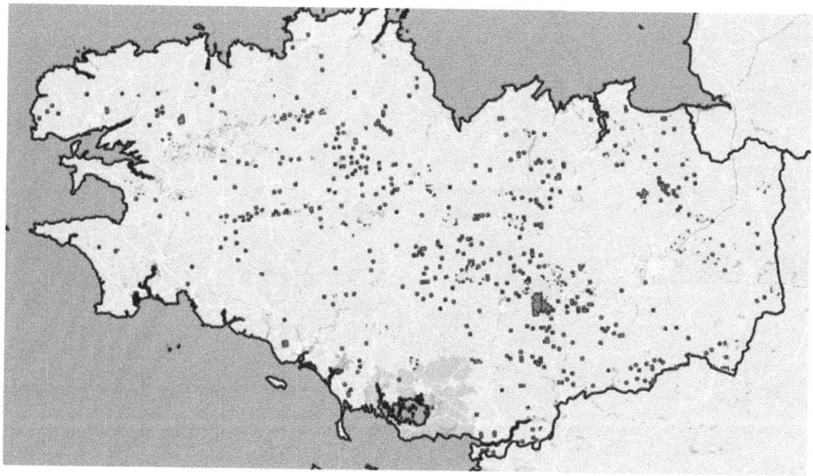

Figure 5.13. *Mapping result of the difference of protected areas. For a color version of the figure, see www.iste.co.uk/baghdadi/qgis3.zip*

5.2.5. *Step 5: consideration of regional wind policy and wind energy criteria*

The model of this fifth step adds a set of criteria relating to wind energy in Brittany to the analysis (wind development zones, regional wind scale and wind power density) as illustrated in Figure 5.14. Figure 5.15 shows the result of this step.

5.2.5.1. *Inclusion into the regional wind energy plan*

The installation of the new wind farm must be within the scope of the regional wind energy plan (which excludes certain restricted areas such as aerodromes and other sensitive areas). We need to apply a difference to exclude any potentially significant areas that are not within the regulatory boundaries (it is also possible to apply a clip).

5.2.5.2. *Distance to existing wind farms*

The idea here is to take into account a minimum distance to existing wind farms. The aim is to create a fixed 3 km buffer distance on all existing wind farms and to apply a difference in order to exclude these areas.

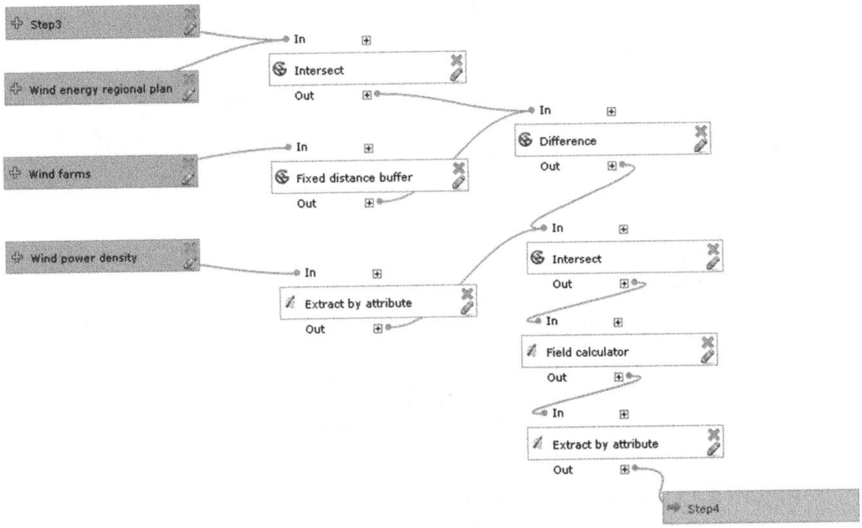

Figure 5.14. *Workflow for consideration of regional wind policy and wind energy potentialities*

5.2.5.3. *Wind power density*

The last criterion linked to the wind is the taking into account of wind power density (up to 80 m)[8]. The aim here is to only keep areas with a relevant wind potential (extraction by attributes based on a minimum of 350 W/m^2) and to find the potential zones of interest located in these zones with high wind potential (clip). In addition, it is necessary to add a surface criterion to keep only the zones of more than 30 ha in order to eliminate the residues of the previously processed geoprocessing processes.

> **QGIS functionalities:**
> - Intersection: *QGIS geoalgorithms > Vector overlay tools > Intersection* **OR** *SAGA > vector polygon Tools > Intersection*
> - Buffer: *QGIS geoalgorithms > Vector geometry tools > Fixed distance buffer* **OR** *SAGA > Vector general tools > Fixed distance buffer*
> - Difference: *QGIS geoalgorithms > Vector overlay tools > Difference* **OR** *SAGA > vector polygon Tools > Difference*

8 The private company EED (*Espace éolien développement*) was commissioned by the Brittany Region to build this database in the context of the regional wind energy plan published in July 2006. The unit used in this dataset is the W/m^2.

- Update area: *QGIS geoalgorithms > Vector table tools > Export / Add geometry columns*
- Select features by area: *QGIS geoalgorithms > Vector selection tools > Extract by attribute*

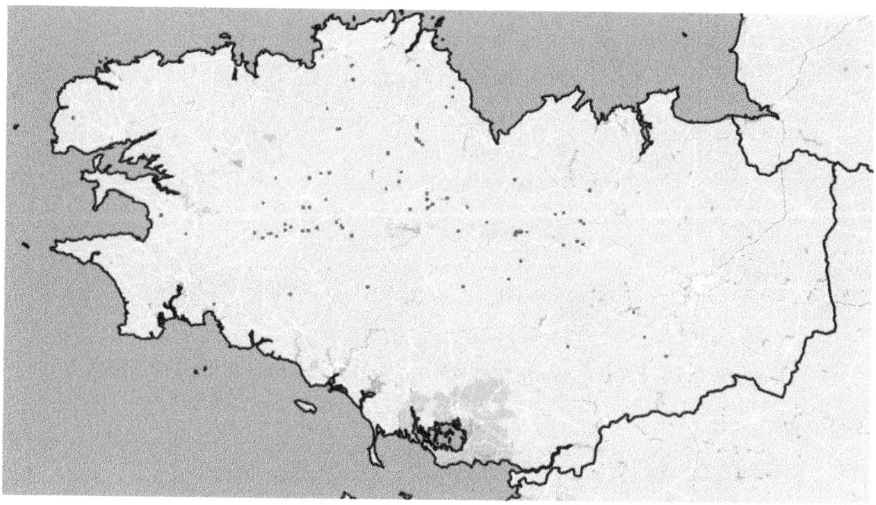

Figure 5.15. *Mapping result of the workflow. For a color version of the figure, see www.iste.co.uk/baghdadi/qgis3.zip*

5.2.6. *Step 6: proximity to power lines*

The last step of the analysis is to take into account the existing power lines. The idea is to install the new wind farm near power lines in order to avoid building a new distribution network. This step, illustrated in Figure 5.16, is structured into three substeps: (1) apply a 1 km buffer around power lines; (2) extract the zones in contact with these buffers by location; (3) keep only areas with a surface area greater than 1 km^2 (area relevant to a medium-sized park).

The three optimal wind farm locations identified by the model are shown in green in Figure 5.17. It is appropriate to finalize the analysis to verify what these areas look like at the local level by zooming and adding satellite imagery. This visual control allows us to validate or invalidate the results of the model by checking whether the zones present incompatibilities for the installation of a wind farm (presence of houses, forest, water bodies, etc.).

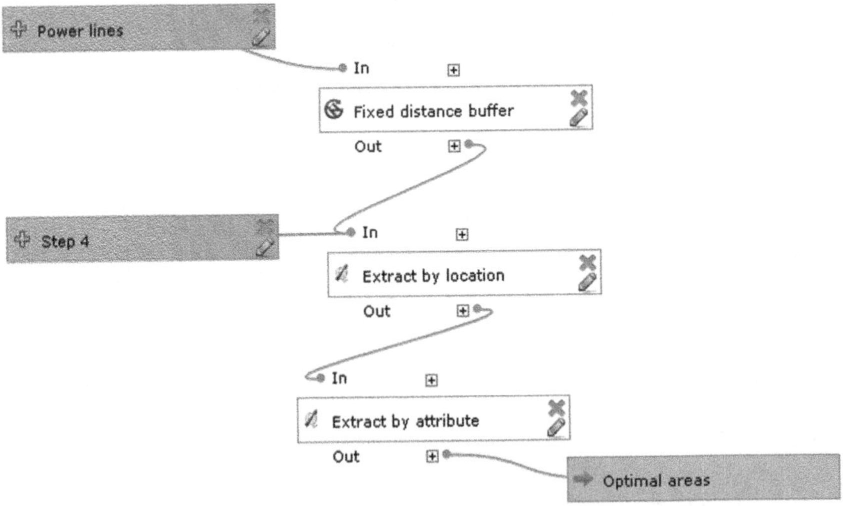

Figure 5.16. *Workflow for consideration of proximity to power lines*

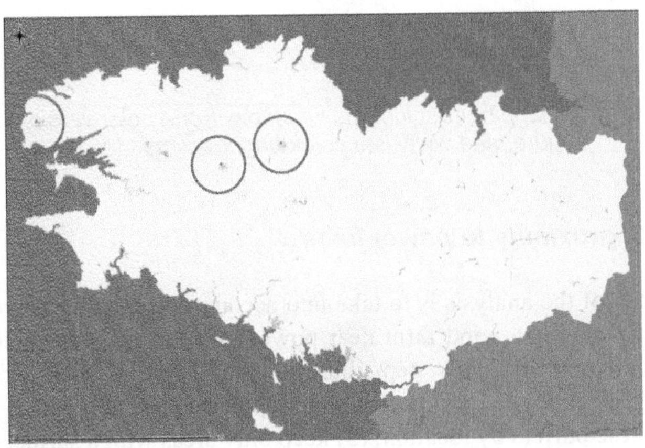

Figure 5.17. *Map of the three optimal areas for installation of new wind farm*

As illustrated in Figure 5.18, the three identified areas do not contain dwellings. However, zones b and c are located near houses. Concerning land use, the zone a contains wooded areas. To compile this study, it would be possible to add other criteria and information to refine the model (cadaster, zoning, flood risks, etc.).

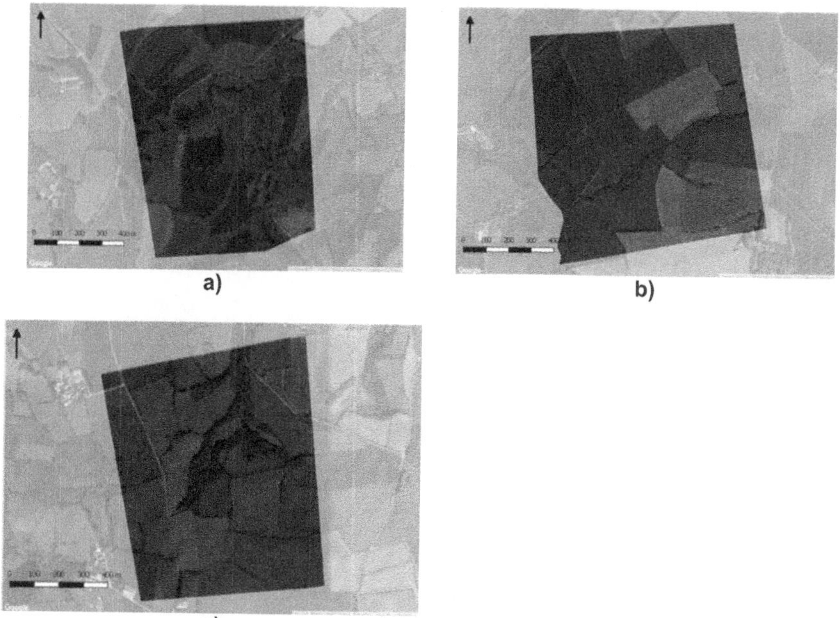

Figure 5.18. *Zoom on three optimal areas with satellite imagery*

QGIS functionalities:

- Buffer: *QGIS geoalgorithms > Vector geometry tools > Fixed distance buffer* **OR** *SAGA > Vector general tools > Fixed distance buffer*

- Select features by location: *QGIS geoalgorithms > Vector selection tools > Extract by location*

- Select features by area: *QGIS geoalgorithms > Vector selection tools > Extract by attribute*

Throughout the chapter, for both ease of understanding and for the stability of the graphic modeler, the various workflows have been built and executed in an isolated way. However, it is possible, once a model is stabilized, to perform all the steps at once by concatenating workflows in a single workflow. It suffices to add the inputs and to integrate the various workflows into the model (Figure 5.19). However, it is advisable for a first use of the graphic modeler to go step by step as when starting to query in SQL.

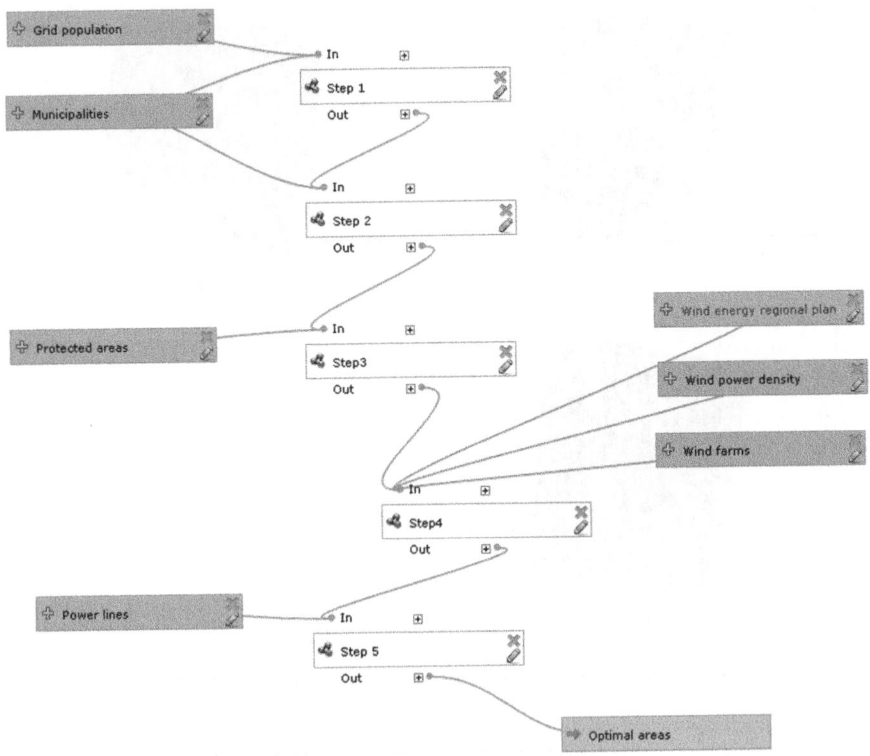

Figure 5.19. *Workflow concatenation*

5.3. Implementation of the application

This section presents the implementation of workflows into QGIS automation for the installation of a wind farm in Brittany.

5.3.1. *Software and data*

5.3.1.1. *Software required*

The process discussed in this chapter uses the basic functionality of the QGIS software (version 2.18). It is also necessary to install some additional extensions to carry out the processing chain: *QuickOSM* and *OpenLayersPlugin*.

5.3.1.2. *Data*

– **Population grid**

Name of layer: Fichier métropole – La base – Métropole

Link: https://www.insee.fr/fr/statistiques/1405815

– **Municipalities of Brittany**

Name of layer: l_communes_2016

WFS URL: http://geobretagne.fr/geoserver/dreal_b/wms

– **Wind farms**

Name of layer: Parcs éoliens en Bretagne

WFS URL: http://geobretagne.fr/geoserver/dreal_b/wms

– **Wind energy regional plan**

Name of layer: Schéma Régional Eolien en Bretagne

WFS URL: http://geobretagne.fr/geoserver/dreal_b/wms

– **Wind power density**

Name of layer: Vents – Densité de puissance (altitude de référence 80 m)

WFS URL: http://ows.region-bretagne.fr/geoserver/rb/wfs

– **Regional natural park**

Name of layer: Parc Naturel Régional

WFS URL: http://ws.carmen.application.developpement-durable.gouv.fr/WFS/10/Nature_Paysage

– **Important bird areas**

Name of layer: ZICO

WFS URL : http://ws.carmen.application.developpement-durable.gouv.fr/WFS/10/Nature_Paysage

– **ZNIEFF 1**

Name of layer: ZNIEFF 1

WFS URL: http://ws.carmen.application.developpement-durable.gouv.fr/WFS/10/Nature_Paysage

– **ZNIEFF 2**

Name of layer: ZNIEFF 2

WFS URL: http://ws.carmen.application.developpement-durable.gouv.fr/WFS/10/Nature_Paysage

– Biotope protected area

Name of layer: Arrêtées de biotope polygones

WFS URL: http://ws.carmen.application.developpement-durable.gouv.fr/WFS/10/Nature_Paysage

– Regional nature reserve

Name of layer: Réserve naturelle régionale

WFS URL: http://ws.carmen.application.developpement-durable.gouv.fr/WFS/10/Nature_Paysage

– National nature reserve

Name of layer: Réserve naturelle nationale

WFS URL: http://ws.carmen.application.developpement-durable.gouv.fr/WFS/10/Nature_Paysage

– World heritage

Name of layer: Patrimoine mondial

WFS URL: http://ws.carmen.application.developpement-durable.gouv.fr/WFS/10/Nature_Paysage

– Registered sites

Name of layer: Sites inscrits (surfacique)

WFS URL: http://ws.carmen.application.developpement-durable.gouv.fr/WFS/10/Nature_Paysage

– Classified sites

Name of layer: Sites classés (surfacique)

WFS URL: http://ws.carmen.application.developpement-durable.gouv.fr/WFS/10/Nature_Paysage

– Habitat directive

Name of layer: Directive habitats (ZSC, SIC, pSIC)

WFS URL: http://ws.carmen.application.developpement-durable.gouv.fr/WFS/10/Nature_Paysage

– Birds directive

Name of layer: Directive oiseaux (ZPS)

WFS URL: http://ws.carmen.application.developpement-durable.gouv.fr/WFS/10/Nature_Paysage

5.3.2. *Step 1: downloading datasets*

This first step consists of downloading the datasets that will be mobilized in the implementation of the different automated workflows.

5.3.2.1. *Download data with Britanny Region WFS*

<table>
<tr><th>Process</th><th>Handling in QGIS</th></tr>
<tr><td>1. Add a WFS connection</td><td>In QGIS:
• Click on Add WFS 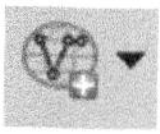
In the Dialog box:
• Create a new connection (New)
• Add Connection name
• Add URL connection
http://ows.region-bretagne.fr/geoserver/rb/wfs
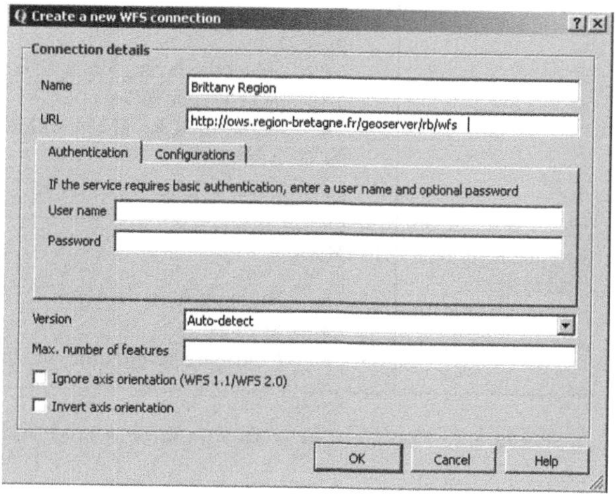

• Click on OK
• Click on connect
→ The different layers of the WFS are accessible</td></tr>
</table>

	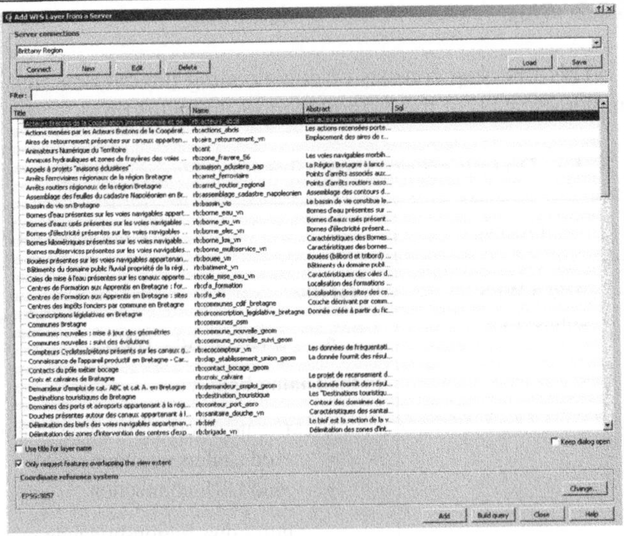
2. Download layer	In the **WFS connection window**: • Select the layer : *Vents – Densité de puissance (altitude de référence 80 m)* • Check the CRS (2154 – Lambert 93) • Add layer • Layer is loaded into QGIS
3. Save the layer on local system	In **QGIS**: • Right click on the layer • Save as… • Configure save (name, file directory)

Table 5.2. *Steps for downloading data from the Britanny Region WFS*

5.3.2.2. *Download data with DREAL WFS*

Process	Handling in QGIS
1. Add a WFS connection	In **QGIS**:  • Click on Add **WFS** In the **dialog box**: • Create a new connection • Add Connection name

	• Add URL connection http://geobretagne.fr/geoserver/dreal_b/wms • Click on OK • Click on connect
1. Download datasets	In the **WFS connection window**: • Select *l_communes_2016 layer* • Check the CRS (2154 – Lambert 93) • Add layer • The layer is loaded into QGIS **Repeat the process for the other layers:** • *Wind farms* • *Wind energy regional plan*
2. Save datasets on local system	In **QGIS**: • Right click on layer • Save as • Configure save (name, file directory)

Table 5.3. *Steps for downloading data from the Britanny DREAL WFS*

5.3.2.3. *Download data with CARMEN WFS*

Process	Handling in QGIS
1. Add a WFS connection	In **QGIS**: 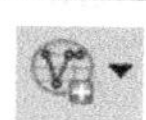 • Click on Add **WFS** In the **dialog box**: • Create a new connection • Add Connection name • Add URL connection http://ws.carmen.application.developpement-durable.gouv.fr/WFS/10/Nature_Paysage? • Click on OK • Click on Connect
2. Download datasets	In **WFS connection window**: • Select *Naturel Régional* layer

	• Check the CRS (2154 – Lambert 93) • Add layer **Repeat the process for the other layers:** • *ZICO* • *ZNIEFF 1* • *ZNIEFF 2* • *Arrêtées de biotope polygones* • *Réserve naturelle régionale* • *Réserve naturelle nationale* • *Patrimoine mondial* • *Sites inscrits* • *Sites classés* • *Directive habitats (ZSC, SIC, pSIC)* • *Directive oiseaux (ZPS)*
3. Save datasets on local system	In **QGIS**: • Right click on layer • Save as • Configure save (name, file directory)

Table 5.4. *Steps for downloading data from CARMEN WFS*

5.3.2.4. *Download data with QuickOSM*

Process	Handling in QGIS
1. Configure query for download of Power lines of Brittany	In **QGIS**: 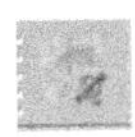 • Launch QuickOSM extension In **Quick query menu**: • Configure Key: *Power* • Configure Value: *Line* • Choose as extent *Municipalities* layer

<table>
<tr><td></td><td>
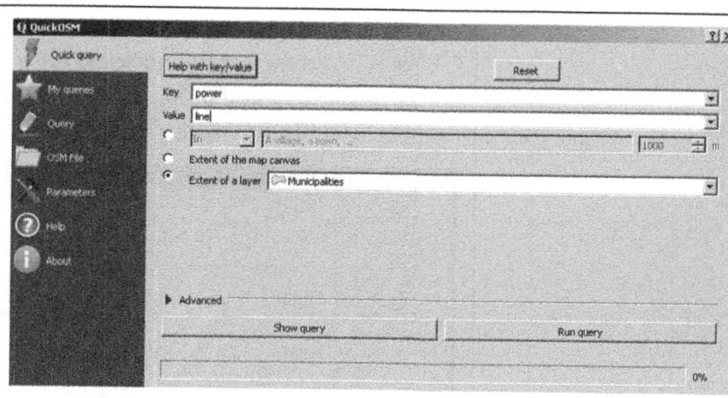

<ul><li>Click on Run query</li></ul>

→ The Power lines layer has downloaded into QGIS in WGS84 and is stored in the cache</td></tr>
<tr><td>2. Save and reproject layer</td><td>In QGIS:
<ul><li>Right click on layer</li><li>Save as</li><li>Configure save (name, file directory, CRS → EPSG 2154 – Lambert 93)</li></ul></td></tr>
</table>

Table 5.5. *Steps for downloading data with QuickOSM*

5.3.3. *Step 2: preparation of population grid dataset*

<table>
<tr><th>Process</th><th>Handling in QGIS</th></tr>
<tr><td>1. Run Graphical Modeler</td><td>In QGIS:
<ul><li>Processing > Graphical Modeler</li></ul>
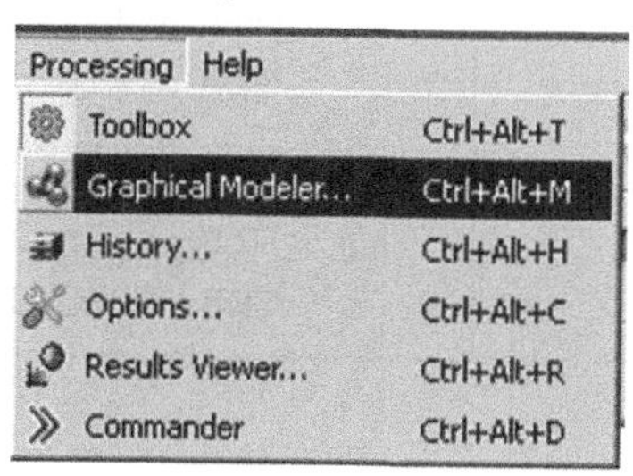
</td></tr>
<tr><td>2. Define model inputs</td><td>In Graphical modeler interface:
<ul><li>Add the first vector layer (double-click on its name or drag and drop) to the modeler canvas</li></ul></td></tr>
</table>

<table>
<tr>
<td></td>
<td>

- Configure name (*Population Grid*)
- Add the second vector layer (double-click on its name or drag and drop) to the modeler canvas
- Configure name (*Municipalities*)

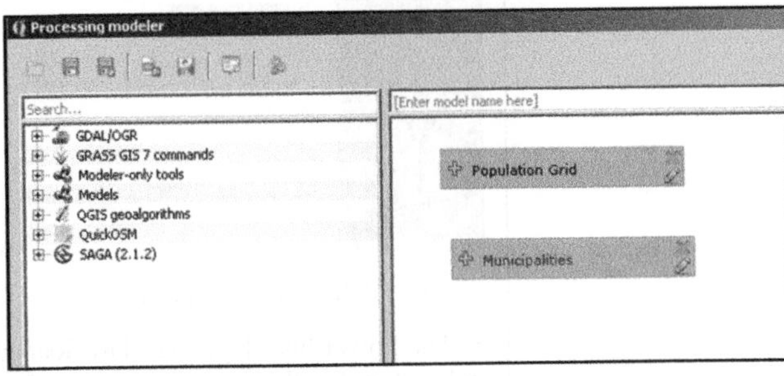

</td>
</tr>
<tr>
<td>3. Add and configure Reproject layer algorithms</td>
<td>

In **Graphical modeler interface**:

- Open Algorithms tab

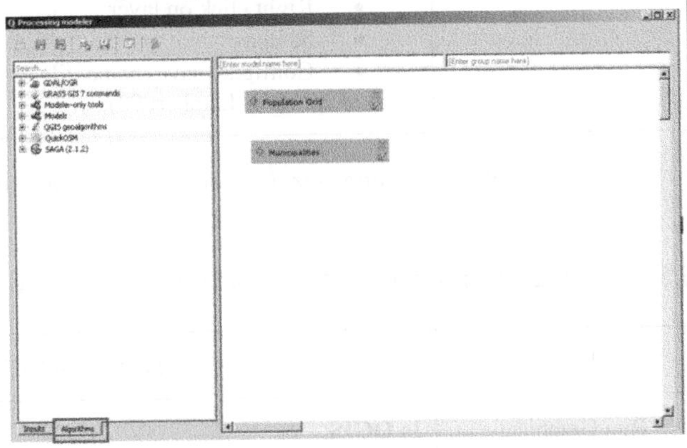

- Add **Reproject layer** algorithm (double-click on its name or drag and drop) to reproject the *Population grid* layer

QGIS geoalgorithms > Vector general tools > Reproject layer

- Configure algorithm parameters (Input layer and Target CRS – EPSG 2154)

</td>
</tr>
</table>

<table>
<tr>
<td></td>
<td>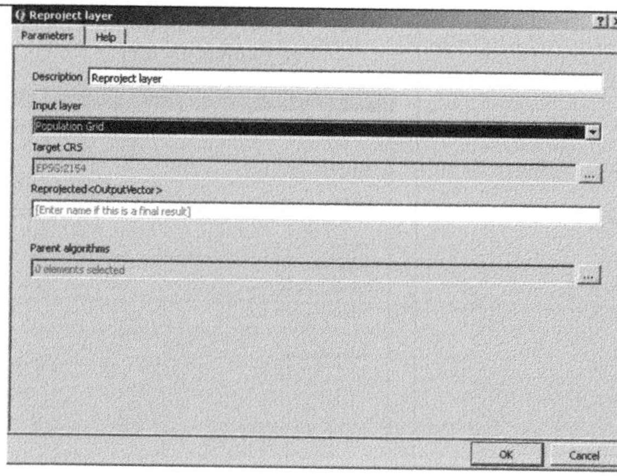
</td>
</tr>
<tr>
<td>4. Add and configure Convert format algorithm</td>
<td>In Graphical modeler interface:
<ul>
<li>Add the Convert format algorithm to convert the Population grid layer (MapInfo MIF) into ESRI Shapefile

Library GDAL/OGR > Conversion > Convert</li>
<li>Configure algorithm parameters (Input layer and Destination Format → ESRI Shapefile)</li>
</ul>
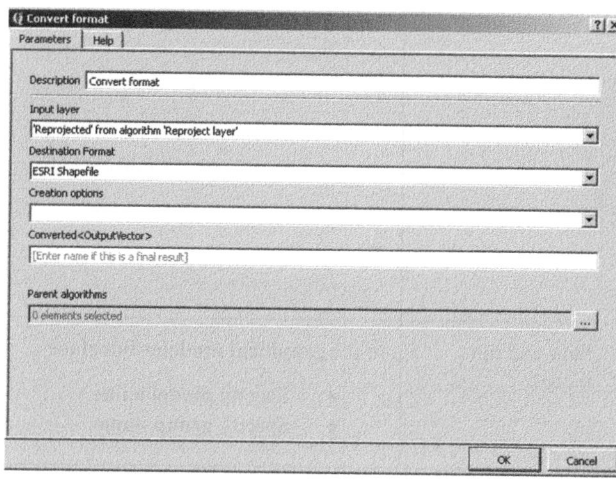
</td>
</tr>
<tr>
<td>5. Add and configure Clip algorithm</td>
<td>In the Graphical modeler interface:
<ul>
<li>Add the Clip algorithm to clip the Grid population layer by the Municipalities layer</li>
</ul>
QGIS Geoalgorithms > Vector overlay tools > Clip</td>
</tr>
</table>

<table>
<tr>
<td></td>
<td>

- Configure algorithm (Input layer, Clip layer and name of Output vector)

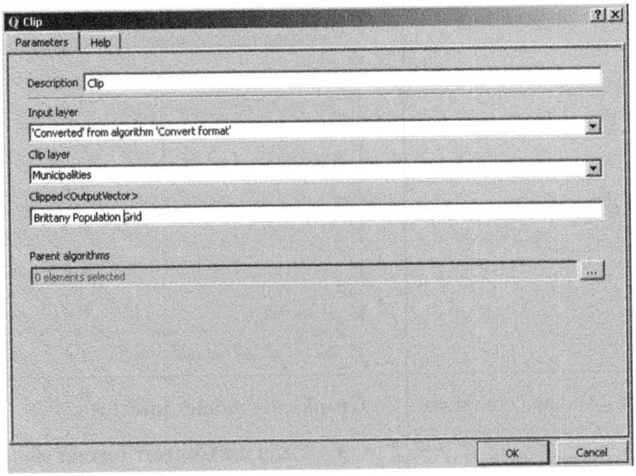

→ Overview of workflow

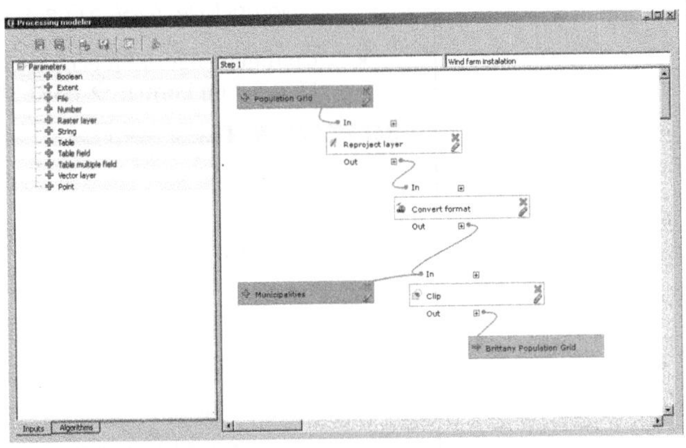

</td>
</tr>
<tr>
<td>6. Save and run model</td>
<td>

In the **graphical modeler interface**:

- Specify **model name**
- Specify **group name**

</td>
</tr>
</table>

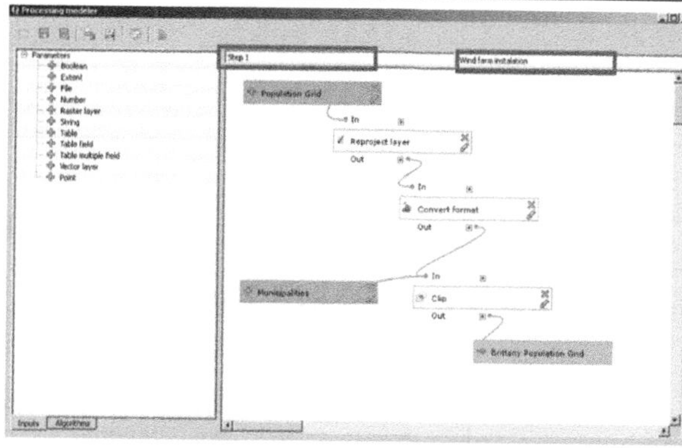

- Save model
- Run model

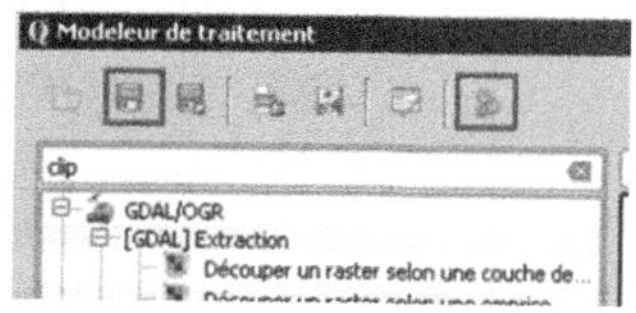

- Define the file path of Inputs
- Run model

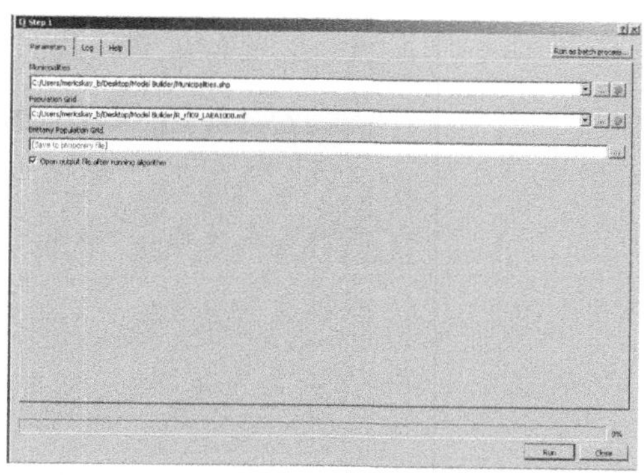

- Once the model runs, the new layer (result of the first workflow) loads into QGIS

Table 5.6. *Steps for the preparation of population grid dataset*

5.3.4. *Step 3: identification of inhabited areas*

Process	Handling in QGIS
1. Run Graphical Modeler	In **QGIS**: • Processing > Graphical Modeler
2. Add model inputs	In the **graphical modeler interface**: • Add the inputs of the models: o *Grid population* o *Municipalities*
3. Add and configure *Symmetrical difference* algorithm	In the **graphical modeler interface**: • Add the **Symmetrical difference** algorithm *SAGA > Vector polygon tools > Symmetrical difference* • Configure algorithm (Input layer, Difference symmetrical layer) 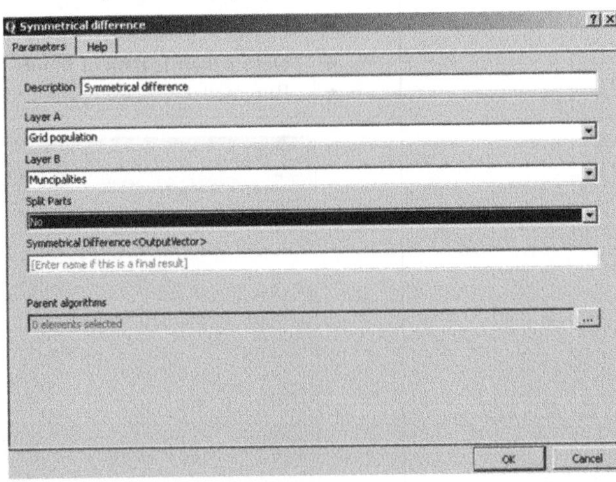
4. Add and configure the *Difference* algorithm	In the **graphical modeler interface**: • Add **Difference** algorithm *SAGA > Vector polygon tools > Difference* • Configure algorithm (Input layer, Difference layer)

<table>
<tr>
<td></td>
<td>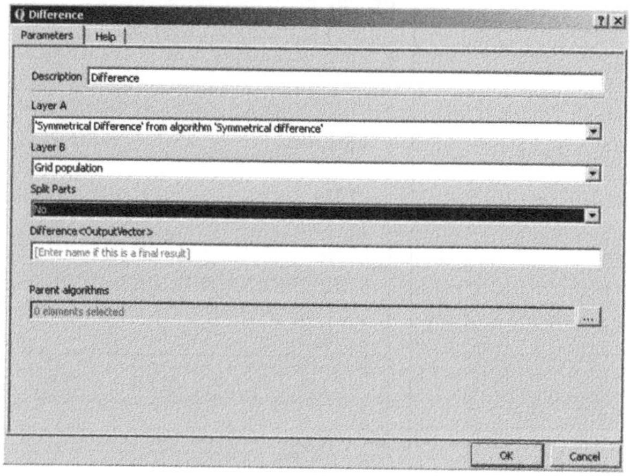
</td>
</tr>
<tr>
<td>5. Add and configure the Add geometry columns algorithm</td>
<td>In the graphical modeler interface:
<ul><li>Add Export/Add geometry columns algorithm</li></ul>
QGIS geoalgorithm > Vector table tools > Export/ Add geometry columns
<ul><li>Configure algorithm (Input layer, Calculate Using)</li></ul>
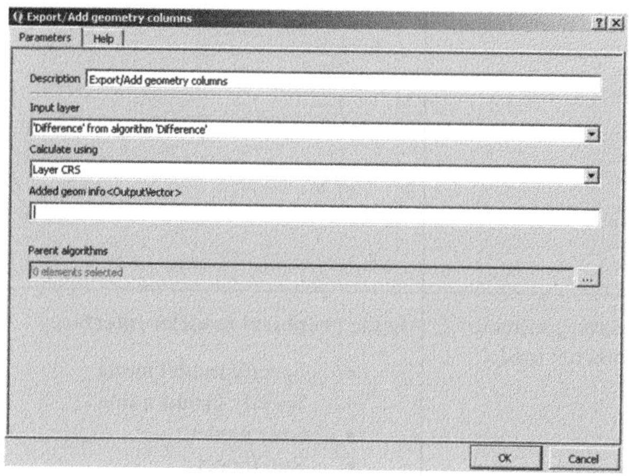
</td>
</tr>
<tr>
<td>6. Add and configure the Extract by attribute algorithm</td>
<td>In the graphical modeler interface:
<ul><li>Add the Extract by attribute algorithm</li></ul>
QGIS geoalgorithm > Vector selection tools > Extract by attribute
<ul><li>Configure algorithm (Input layer, Selection attribute, Operator, Value and name of Output vector)</li></ul></td>
</tr>
</table>

<table>
<tr>
<td></td>
<td>
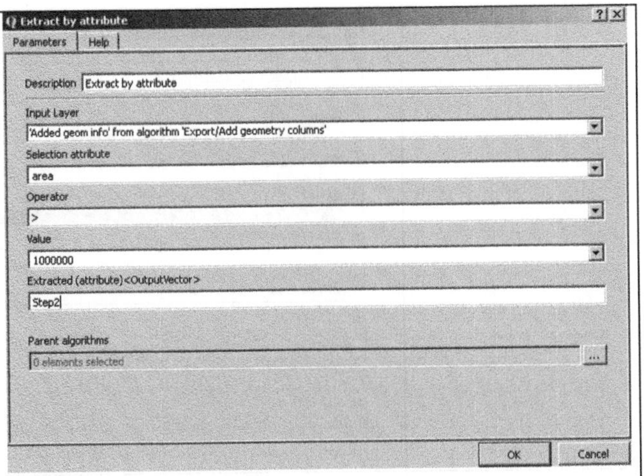

→ Overview of the workflow

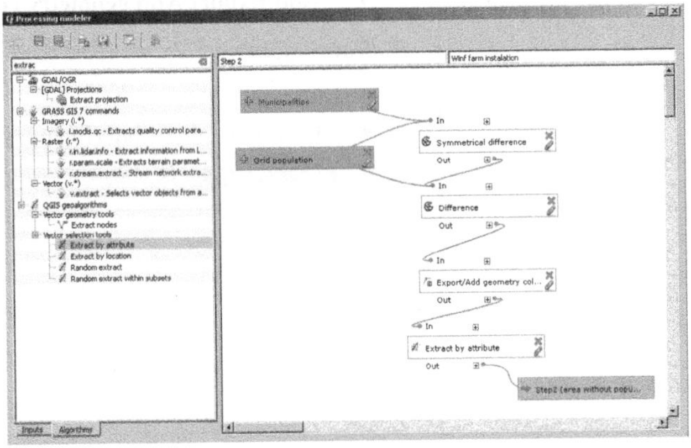

</td>
</tr>
<tr>
<td>7. Save, configure and run model</td>
<td>In the graphical modeler interface:
<ul>
<li>Specify model name</li>
<li>Specify group name</li>
<li>Save model</li>
<li>Run model</li>
<li>Define the file path of Inputs</li>
</ul>
</td>
</tr>
</table>

	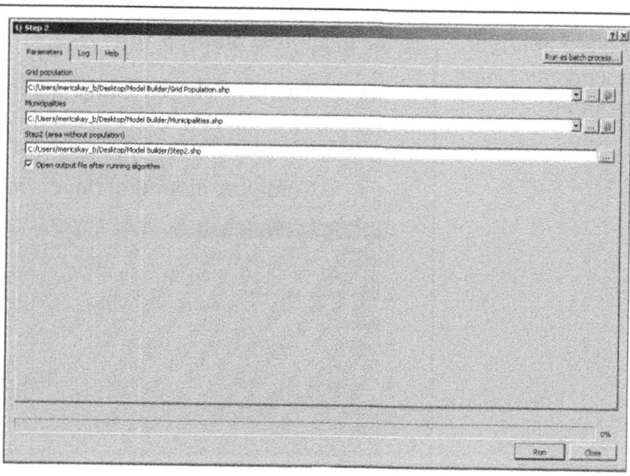 • Run model • Once the model runs, the new layer (result of the workflow) loads into QGIS

Table 5.7. *Steps for the identification of inhabited areas*

5.3.5. *Step 4: consideration of protected areas*

5.3.5.1. *Union of different protected areas*

Process	Handling in QGIS
1. Run Graphical Modeler	In **QGIS**: • Processing > Graphical Modeler
2. Add model inputs	In the **graphical modeler interface**: • Add the inputs of the model o *Regional natural park* o *ZICO* o *ZNIEFF 1* o *ZNIEFF 2* o *Biotope areas* o *Regional nature reserve* o *Regional nature reserve* o *World heritage* o *Registered sites* o *Classified sites* o *Habitat directive* o *Birds directive*

<table>
<tr>
<td>3. Add and configure the Polygon Union algorithm</td>
<td>
In the graphical modeler interface:

- Add the **Polygon Union** algorithm

  SAGA > Vector polygon tools > Polygon Union

- Configure algorithm (layers and Output name)

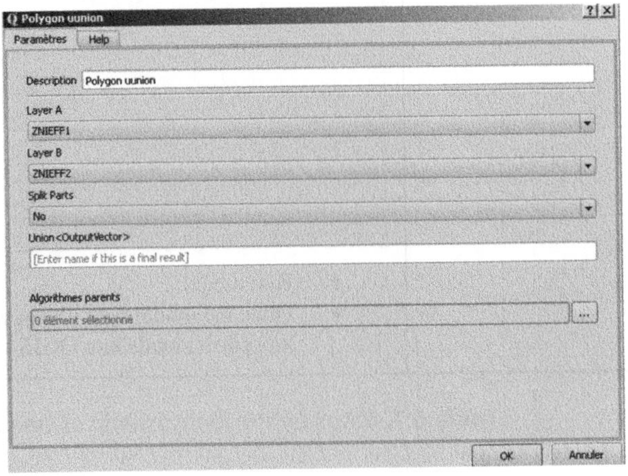

- Repeat this operation 10 times for the union of all inputs of model

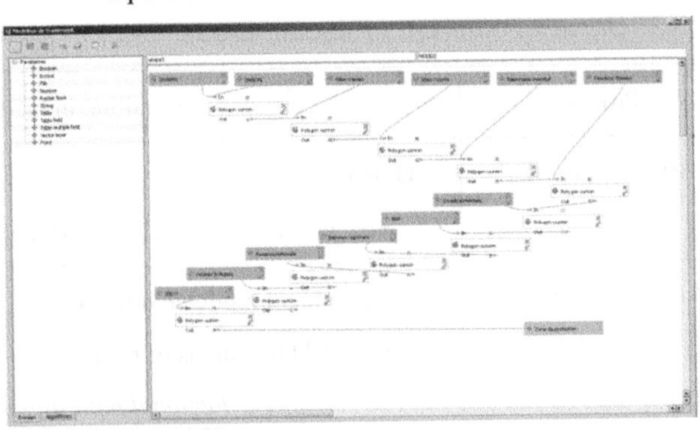
</td>
</tr>
<tr>
<td>4. Configure and run the model</td>
<td>
In the graphical modeler interface:

- Run model
- Define the file path of Inputs
- Run model
- Save this layer
</td>
</tr>
</table>

Table 5.8. *Steps of the union of different protected areas*

5.3.5.2. *Apply a difference between the protected areas union layer and areas without population layer*

Process	Handling in QGIS
1. Run the Graphical Modeler	In **QGIS**: • Processing > Graphical Modeler
2. Add model inputs	In the **graphical modeler interface**: • Add the inputs of the model: o *Protected areas (result of the previous task)* o *Areas without population*
3. Add and configure *Difference* algorithm	In the **graphical modeler interface**: • Add **Difference** algorithm *SAGA > Vector polygon tools > Difference* • Configure algorithm (Input layer, Difference layer) 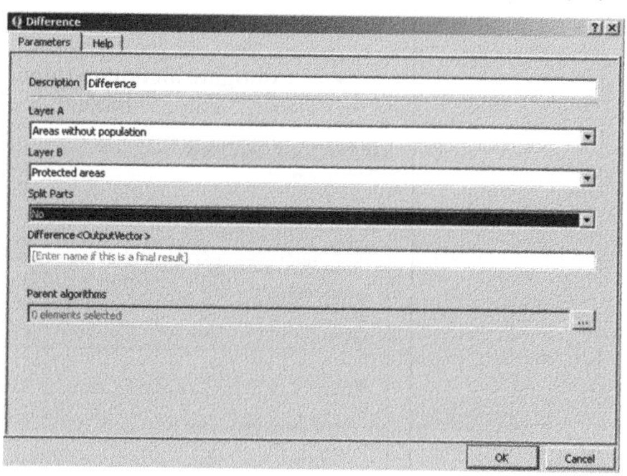
4. Add and configure the *Field Calculator* algorithm	In the **graphical modeler interface**: • Add the **Field Calculator** algorithm *QGIS algorithm > Vector table tools > Field calculator* • Configure algorithm (Input layer, Result field name, Field type, Field length, Create new field and Formula = *$area*)

<table>
<tr><td></td><td>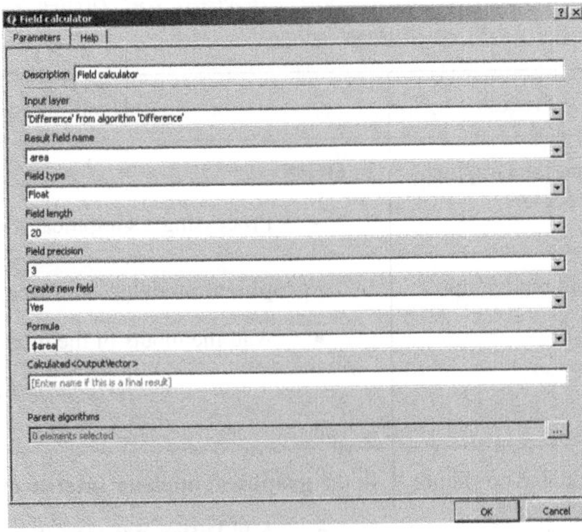
</td></tr>
<tr><td>5. Add and configure the Extract by attribute algorithm</td><td>In the graphical modeler interface:

• Add the Extract by attribute algorithm

QGIS algorithm > Vector selection tools > Extract by attribute

• Configure algorithm (Input layer, Selection attribute = area, Value > 100000, Name of output vector)

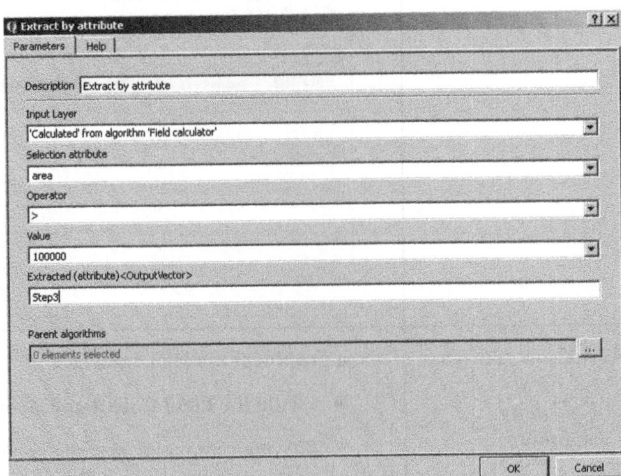

➔ Workflow overview</td></tr>
</table>

	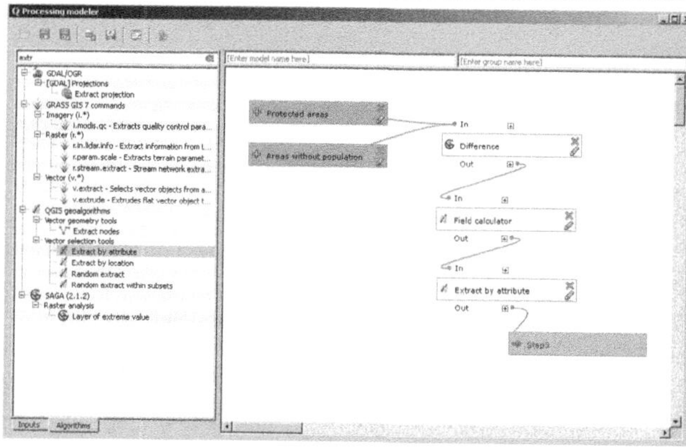
6. Save, configure and run the model	In the **graphical modeler interface**: • Specify **model name** • Specify **group name** • Save model • Run model • Define the file path of Inputs 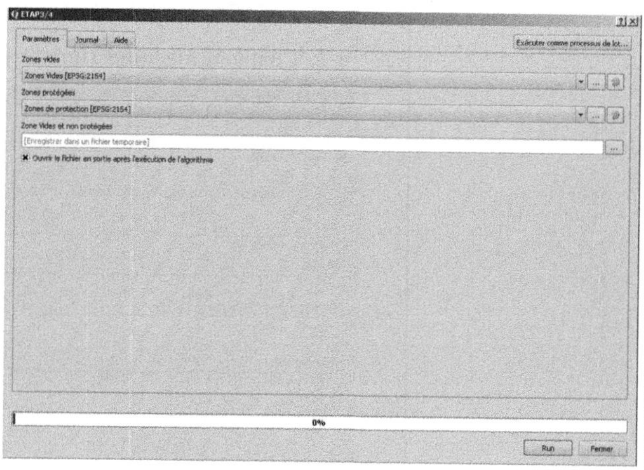 • Run model • Once the model runs the new layer (result of the workflow) loads into QGIS

Table 5.9. *Steps to apply a difference between protected areas union layer and areas without population layer*

5.3.6. *Step 5: consideration of the regional wind policy and wind energy criteria*

<table>
<tr><th>Process</th><th>Handling in QGIS</th></tr>
<tr><td>1. Run the Graphical Modeler</td><td>In QGIS:
• Processing > Graphical Modeler</td></tr>
<tr><td>2. Add model inputs</td><td>In the graphical modeler interface:
• Add the inputs of the model:
o Result of last workflow (areas without population and non-protected)
o Wind farms
o Regional wind energy plan
o Wind power density</td></tr>
<tr><td>3. Add and configure the Intersect algorithm</td><td>In the graphical modeler interface:
• Add Intersect algorithm
SAGA > Vector polygon tools > Intersect
• Configure algorithm (Input layer, Intersect layer)

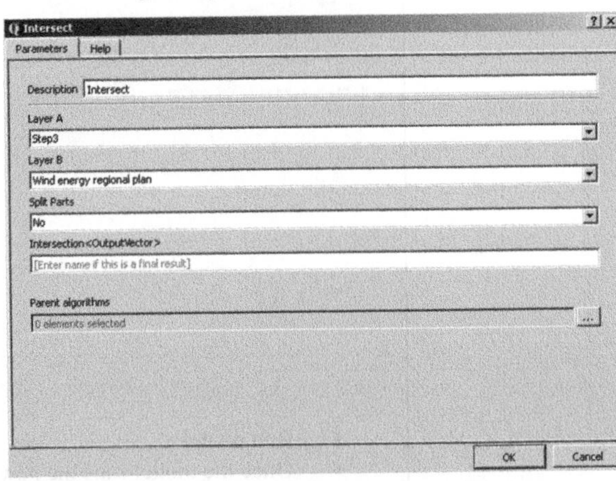

</td></tr>
<tr><td>4. Add and configure the Fixed distance buffer algorithm</td><td>In the graphical modeler interface:
• Add the Fixed distance buffer algorithm
SAGA > Vector general tools > Fixed distance buffer
• Configure algorithm (Buffer of distance = 3000 and Dissolve buffer = YES)</td></tr>
</table>

<table>
<tr>
<td></td>
<td>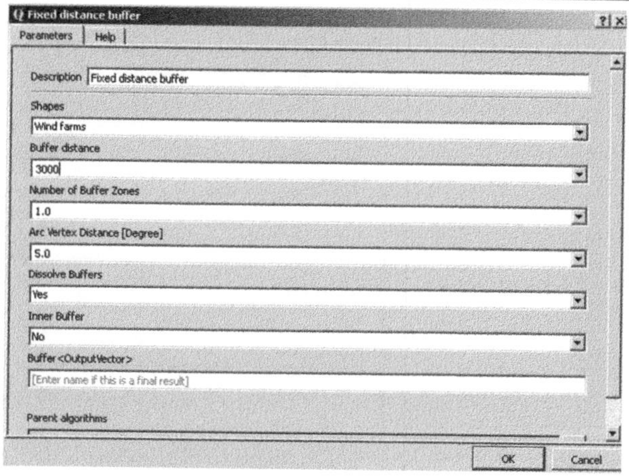
</td>
</tr>
<tr>
<td>5. Add and configure the Difference algorithm</td>
<td>In the graphical modeler interface:
• Add the Difference algorithm
SAGA > Vector polygon tools > Difference
• Configure algorithm (Input layer, Difference layer)
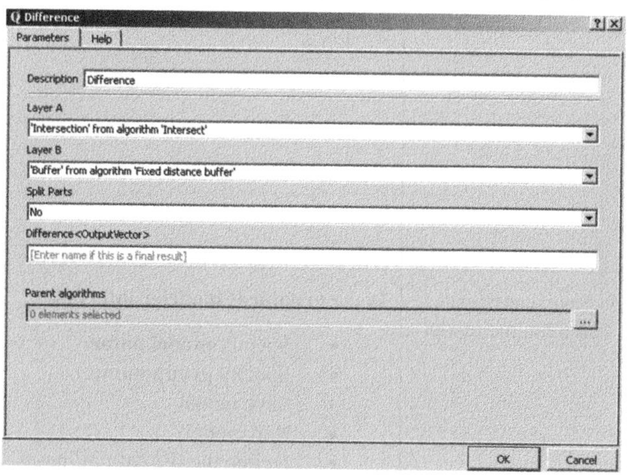
</td>
</tr>
<tr>
<td>6. Add and configure the Extract by attribute algorithm</td>
<td>In the graphical modeler interface:
• Add the Extract by attribute algorithm
QGIS algorithm > Vector selection tools > Extract by attribute
• Configure algorithm (Input layer, Selection attribute = gridcode, Value > 350)</td>
</tr>
</table>

<table>
<tr>
<td></td>
<td>
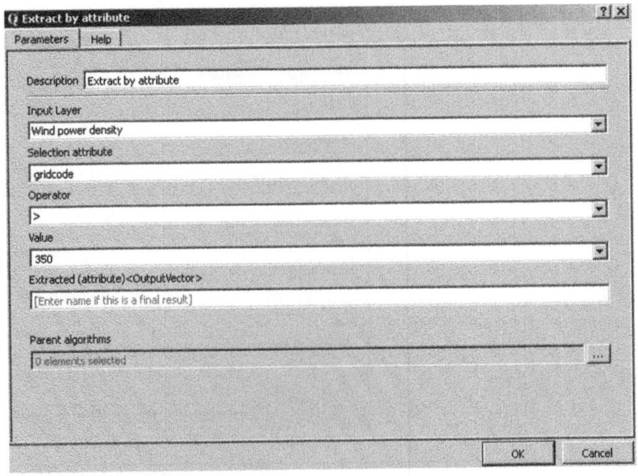

➔ **Workflow overview**

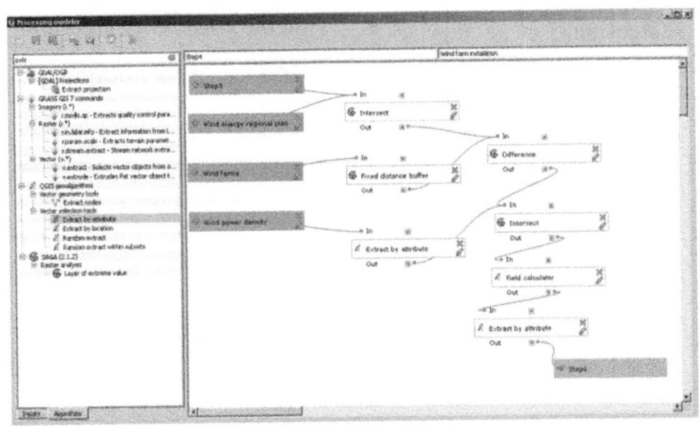
</td>
</tr>
<tr>
<td>7. Save, configure and run the model</td>
<td>In the graphical modeler interface:
<ul>
<li>Specify model name</li>
<li>Specify group name</li>
<li>Save model</li>
<li>Run model</li>
<li>Define the file path of Inputs</li>
</ul>
</td>
</tr>
</table>

	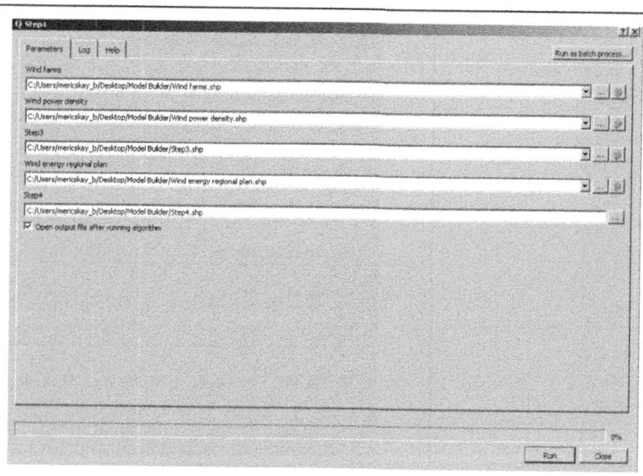 • Run model • Once the model runs, the new layer (result of the workflow) loads into QGIS

Table 5.10. *Steps for the consideration of regional wind policy*

5.3.7. *Step 6: consideration of proximity to power lines*

Process	Handling in QGIS
1. Run the Graphical Modeler	In **QGIS**: • Processing > Graphical Modeler

2. Add model inputs	In the **graphical modeler interface**:
	• Add the inputs of the model
	    o *Result of last workflow*
	    o *Power lines*

3. Add and configure the *Fixed distance buffer* algorithm	In the **graphical modeler interface**: • Add the **Fixed distance buffer** algorithm *SAGA > Vector general tools > Fixed distance buffer* • Configure algorithm (Buffer of distance = 1000 and Dissolve buffer = YES)

<table>
<tr><td></td><td>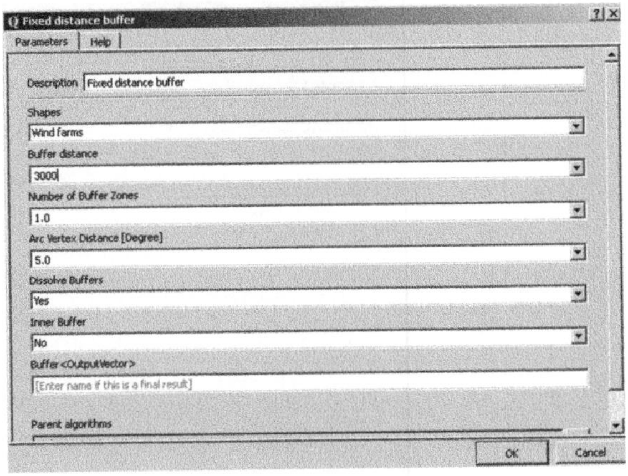
</td></tr>
<tr><td>4. Add and configure the Extract by location algorithm</td><td>In the graphical modeler interface:
• Add the Export by location algorithm
QGIS geoalgorithm > Vector selection tools > Extract by location
• Configure algorithm (Input layer, intersection layer, Geometric predicate = intersects)
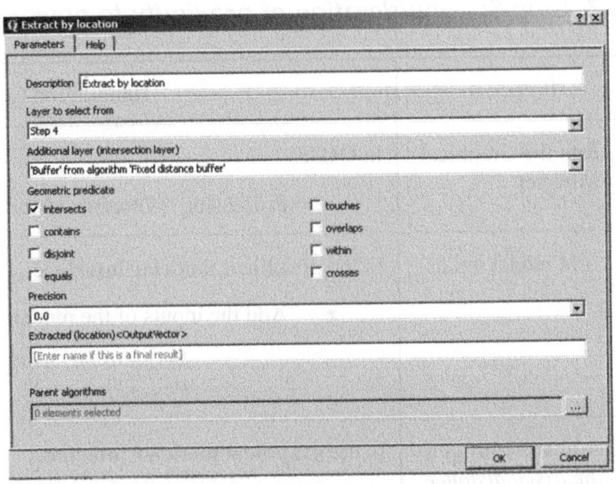
</td></tr>
<tr><td>5. Add and configure the Extract by attribute algorithm</td><td>In the graphical modeler interface:
• Add the Extract by attribute algorithm
QGIS algorithm > Vector selection tools > Extract by attribute
• Configure algorithm (Input layer, Selection attribute = area, Value > 700000)</td></tr>
</table>

<table>
<tr>
<td></td>
<td>
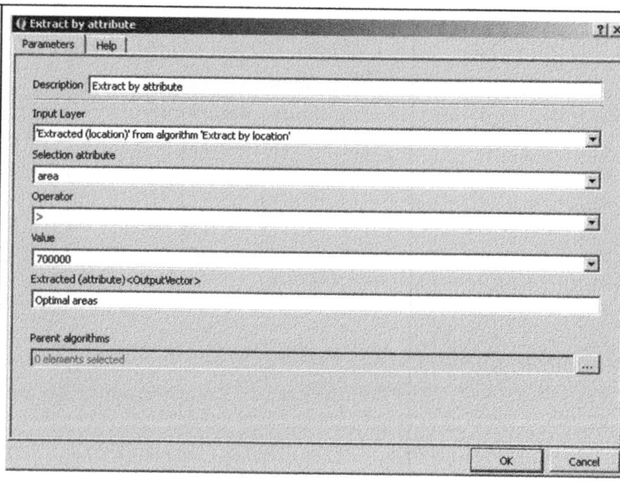

➔ Workflow overview

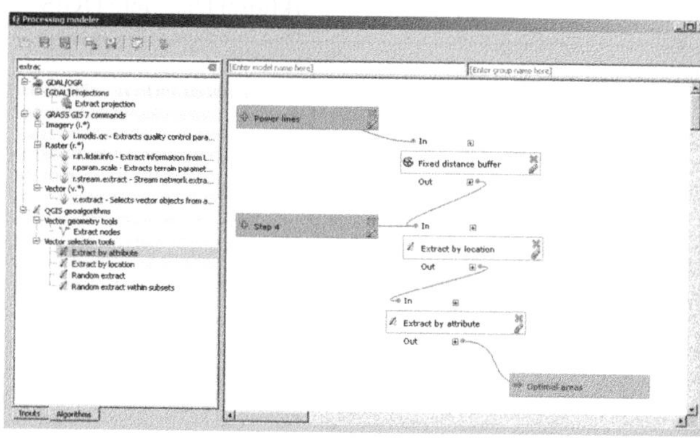

</td>
</tr>
<tr>
<td>6. Save, configure and run the model</td>
<td>
In the graphical modeler interface:

- Specify model name
- Specify group name
- Save model
- Run model
- Define the file path of Inputs
</td>
</tr>
</table>

	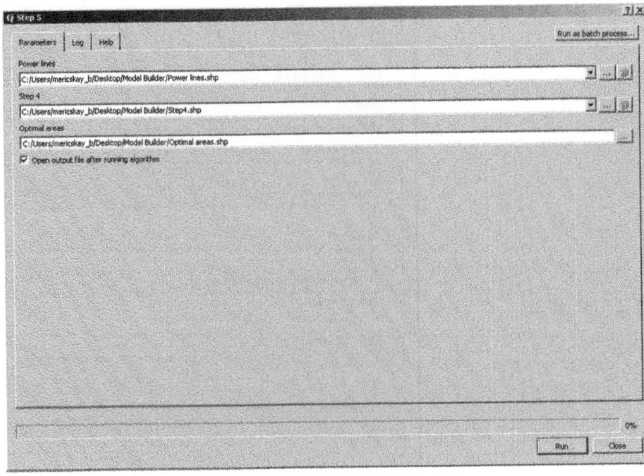 • Run model • Once the model runs the new layer (result of the workflow) loads into QGIS

Table 5.11. *Steps for the consideration of criteria for wind energy*

6

Ecosystemic Services Assessment: Application to Forests for the Preservation of Water Resources in Tropical Islands

6.1. Definition and context

According to the Millenium Ecosystem Assessment, ecosystems provide four main types of services to populations, called ecosystemic services [MIL 03]:

1) the provisioning services (food, fresh water, wood, textiles, fuel, etc.);

2) regulating services allowing an environment where it is physically and biologically possible for humans to live and to be maintained (crop pollination, flood mitigation, climate regulation, etc.);

3) cultural services including recreational, aesthetic, spiritual and educational services;

4) supporting services that represent all the underlying ecosystems' processes for providing the direct services outlined above.

At present, there is no universal indicator to characterize the "health" or "functionality" [MAR 11] of ecosystems. Thus, ecosystems and their respective services are assessed separately according to their own specificities.

In this chapter, we are interested in describing the functioning of forest ecosystems for the protection of water resources within water catchment perimeters (called PPE). The function state assessment is estimated by assessing the loss of services related to forest degradation.

Chapter written by Rémi ANDREOLI and Brice VAN HAAREN.

Degradation of forest ecosystems leads to a loss of associated ecosystemic services protecting water resources and therefore causes:

– a decrease in the abundance and diversity of forest fauna and flora [BRO 02, KLE 89, LAU 98, RAH 01];

– an increase in erosion hazard related to the loss of soil stabilization service [BRA 07B, GYS 05, KEI 03] leading, among other things, to sediment pollution in the rivers;

– loss of flood mitigation service [BRA 07a, BRA 07b BRO 16, FOL 07, GUI 05] and a loss of water provisioning during dry seasons, which can lead to shortages [BRA 07b, SMA 01].

6.2. Method

The methodology presented here is based on the experience gained in New Caledonia (Southwest Pacific Ocean) but it can be broadly applied taking into account the regional specificities of the studied area:

1) the proposed approach evaluates the function state of forest ecosystems based on three criteria:the loss of soil stabilization ecosystemic service;

2) the loss of water regulation ecosystemic service;

3) the loss of resilience service (to bushfires in New Caledonia).

The diagnosis is thus based on an expert decision-tree approach exploiting three parameters that characterize, respectively, the three above-mentioned criteria:

1) soil losses derived from the RUSLE model [AND 18] corresponding to erosion hazard (low, moderate, high, extreme) for each water catchment perimeter;

2) the dominant type of vegetation considered as the “landscape tendency” of each water catchment perimeter (forest, herbaceous and shrub lands, open landscape);

3) the forest fragmentation (low, moderate, high) for each water catchment perimeter.

The method is divided into five main steps detailed in Figure 6.1:

1) PPE (water catchment perimeters) database preparation;

2) erosion hazard determination for each PPE;

3) dominant vegetation type determination for each PPE;

4) forest fragmentation estimates for each PPE;

5) forest function assessment for each PPE.

These steps are performed using several GIS software (QGIS, GRASS) and specific extensions for information correction (Check geometries), perform statistical analysis (Group Stats) and run cross-analysis of vector and matrix data (zonal statistics).

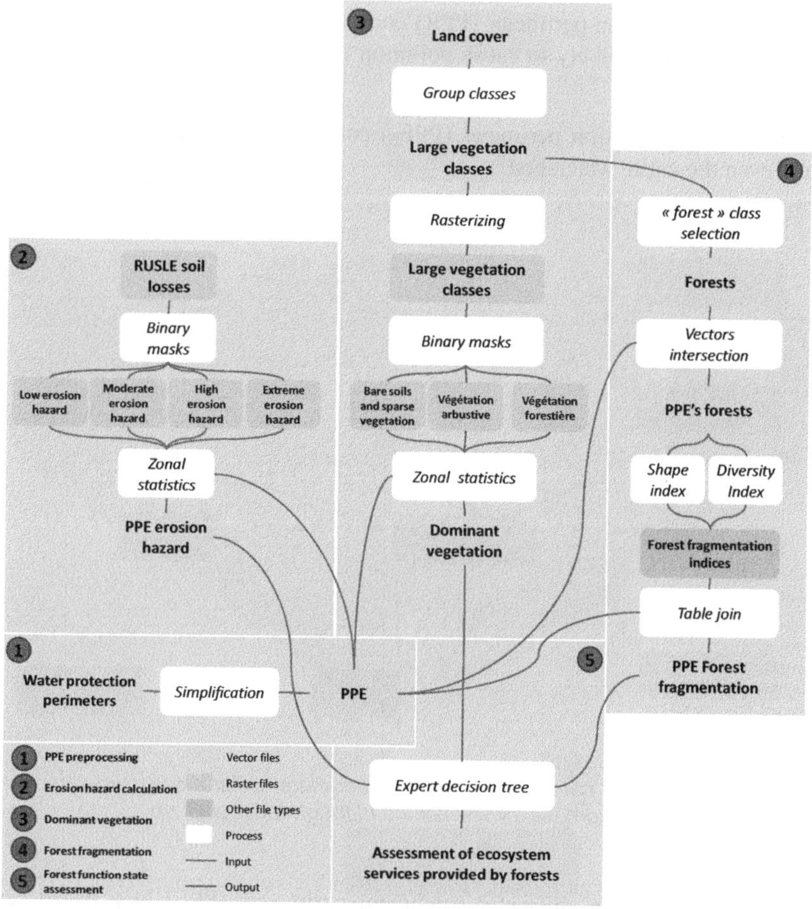

Figure 6.1. *Forest function assessment for water protection within water catchment perimeter algorithm. For a color version of the figure, see www.iste.co.uk/baghdadi/qgis3.zip*

6.2.1. *Water catchment perimeter database (PPE)* preparation

Water catchment perimeters are usually defined around water catchment sites intended for human consumption to ensure the protection of the water resource (Figure 6.2). There are generally three levels of protection:

– the immediate protection perimeter (PPI) that corresponds to the fenced-in site;

– the close protection perimeter (PPR) corresponds to the area of a few hectares for which any activity likely to cause pollution is prohibited or is subject to special prescription;

– the remote protection perimeter (PPE_S) corresponds to the water catchment's supply, even the entire watershed.

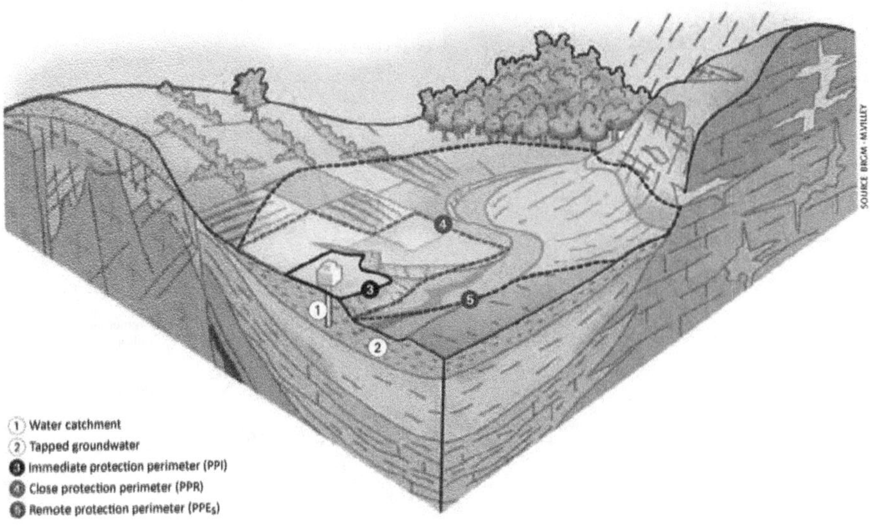

Figure 6.2. *Water catchment protection perimeters. For a color version of the figure, see www.iste.co.uk/baghdadi/qgis3.zip*

The analysis focuses on water catchment perimeters (PPE) in a broad sense, corresponding to the expanded perimeter that includes immediate, close and remote protection perimeters. Indeed, water catchment protection perimeters (PPE_S, PPR and PPI) are geographically defined as distinct polygons based on the spaghetti model (without topological rules and with polygons overlapping). This results in complex overlays and overlaps of geometries. Preparatory work on water catchment perimeters is therefore required.

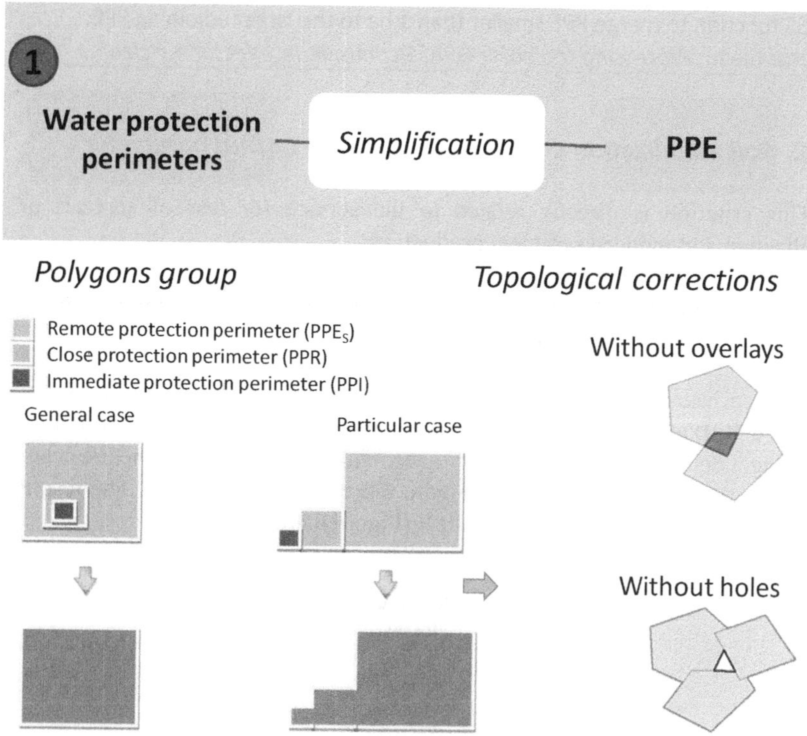

Figure 6.3. *PPE database preparation. For a color version of the figure, see www.iste.co.uk/baghdadi/qgis3.zip*

Perimeters are simplified that includes the following steps (Figure 6.3):

– PPI, PPR and PPE$_S$ merging in continuous and homogeneous unit polygons. These new polygons are named PPE;

QGIS function to group polygons:
- Dissolve: *Vector > Geoprocessing tools > Dissolve...*

– topological correction of the PPE by removing both polygon overlaps and holes inside polygons;

QGIS function to analyze and correct PPE:
- Check geometries: *Vector > Geometry tools > Check geometries*

– PPE smaller than 1 hectare and contiguous to larger PPE are merged into the larger adjoining PPE.

GRASS function to merge PPE smaller than 1 ha to the larger adjoining PPE:
- Vector Clean: *Processing toolbox > GRASS commands > Vector > v.clean...*

6.2.2. *Soil stabilization criterion: the erosion hazard parameter*

This criterion is directly related to the service (or loss of service) of soil stabilization and induced sediment production.

The sediment production is estimated as soil losses due to rainfall and runoff using the RUSLE equation [AND 17].

Erosion hazard assessment for each PPE consists of integrating the soil losses (Figure 6.4(a)) at the spatial unit scale defined by the PPE. This integration is carried out using a decision rule that takes into account the percentage of the area classified according to the erosion intensity (Figure 6.4(b)) [LE 02, LE 04]. The result is an erosion hazard classification of each PPE (Figure 6.4(c)).

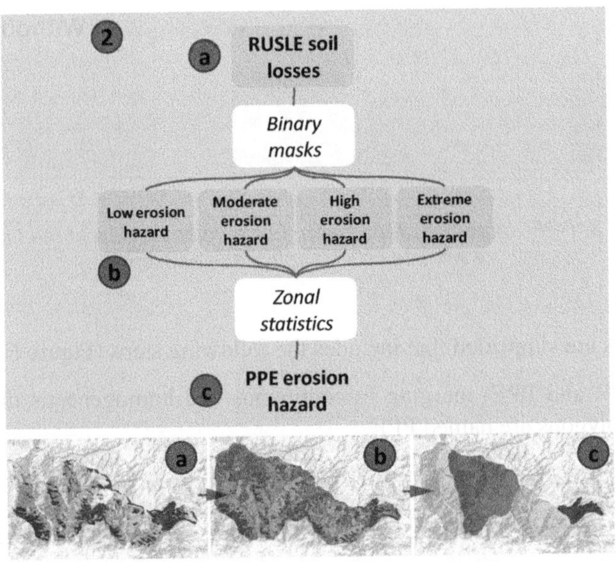

Figure 6.4. *Erosion hazard assessment for each PPE: a) soil loss estimates from RUSLE equation in t/ha/year (from white, values near 0 to red, high soil loss estimates); b) erosion hazard map per pixel according to soil loss estimates: low (green), moderate (yellow), high (orange), extreme (red); c) PPE erosion hazard classification (yellow, low to moderate; orange, moderate to high; red, high to extreme). For a color version of the figure, see www.iste.co.uk/baghdadi/qgis3.zip*

This approach requires several steps:

– define to which erosion hazard class (low, moderate, high, extreme) each pixel belongs according its soil loss value estimated using the RUSLE equation. The result of this step is a map of the whole area at the pixel scale featuring four erosion hazard classes;

– estimate for each PPE:

- the percentage of each erosion hazard class;

- analyze the distribution of each erosion hazard class to characterize the global erosion hazard at the PPE scale according to a defined decision rule.

6.2.2.1. *From RUSLE soil losses to erosion hazard classes*

The RUSLE equation estimates the removal of soil particles on each mesh of the model (pixel) due to water processes (rainfalls and runoff). Soil loss is commonly expressed in t/ha/year.

The correlation between soil loss values modeled by the RUSLE equation and erosion hazard intensity is defined by empirical values beyond which the erosion hazard increases. In New Caledonia, these values are based on a literature review and the statistical distribution of soil loss values at the scale of the entire territory [AND 16].

The values used (Table 6.1) correspond to the three quantiles of the statistical distribution of soil loss values modeled by the RUSLE equation. They are consistent with similar values found in the literature [AND 16].

Low erosion hazard	<2 t/ha/year
Moderate erosion hazard	2 t/ha/ year – 23.3 t/ha/year
High erosion hazard	23.3 t/ha/year – 102.1 t/ha/year
Extreme erosion hazard	>102.1 t/ha/year

Table 6.1. *Thresholds for the definition of erosion hazard classes [AND 16]*

These thresholds are used to create a four-class map of erosion hazard from the soil loss values modeled by the RUSLE equation.

QGIS function to create binary mask:
- Thresholding: *Raster > Raster calculator...*

6.2.2.2. *Erosion hazard determination for each PPE*

From the previous map of erosion hazard in four classes, the respective percentages of each hazard class are calculated for each PPE.

QGIS function to calculate zonal statistics:
- Zonal statistics: *Raster > Zonal statistics > Zonal statistics ...*

Then, the global erosive hazard of each PPE is estimated using a decision rule (Table 6.2) taking into account the area percentages of each erosion hazard class in each PPE according to the method proposed by [LE 02].

Erosion hazard class percentage in the PPE	Global erosion hazard class assignment
- Low erosion hazard >22% and extreme erosion hazard = 0%	**Low to moderate erosion hazard**
- Extreme erosion hazard >0% and extreme erosion hazard ≤18% and low erosion hazard ≤22% - Or high erosion hazard >48% and extreme erosion hazard >0% and extreme erosion hazard ≤18% - Or Extreme erosion hazard >18% and high erosion hazard >0 and [high erosion hazard + moderate erosion hazard] > 29% - Or [extreme erosion hazard + high erosion hazard + moderate erosion hazard] > [low erosion hazard and extreme erosion hazard < high erosion hazard + moderate erosion hazard]	**Moderate to high erosion hazard**
- Extreme erosion hazard >50% - Or Extreme erosion hazard >18% and high erosion hazard >48% - Or [extreme erosion hazard + high erosion hazard + moderate erosion hazard] > low erosion hazard and extreme erosion hazard > [high erosion hazard + moderate erosion hazard] - Or [extreme erosion hazard + high erosion hazard + moderate erosion hazard] >79% and moderate erosion hazard <[high erosion hazard + extreme erosion hazard] - Or extreme erosion hazard >18% and high erosion hazard >0 and [high erosion hazard + moderate erosion hazard] <29%	**High to extreme erosion hazard**

Table 6.2. *Definition of the erosive hazard classes assigned to the PPE of New Caledonia [AND16]*

QGIS function to create a new attribute based on an existing one:
- New attribute calculation: *Processing toolbox > QGIS Geoalgorithms > Vector table tools > Field calculator...*

The advantage of this method resides in the global characterization of the erosion hazard at the scale of an administrative unit taking into account the relative presence of each class of erosion intensity. This integration method is therefore less sensitive to extreme values than the use of the mean or median soil losses. This method also better considers high and extreme erosion hazard areas that are not very dominant within the PPE.

6.2.3. *Water regulation criterion and ecosystem degradation: the dominant vegetation parameter*

The dominant vegetation of a given area provides information on the level of degradation of forest ecosystem due to anthropogenic pressures and the associated services or loss of services. This indicator provides a global index of the provision of a set of ecosystem services. In New Caledonia, the characterization of forest degradation is based on the hypothesis of a forest climax when forests were covering the whole territory before the arrival of the first austronesians [CUR 15, JAF 80, MC 99, IBA 13].

To determine the dominant vegetation of each PPE, the proposed method is similar to the method used to estimate erosion hazard. This method consists of applying the following tasks (Figure 6.5):

– grouping the land-use classes into larger characteristic vegetation classes;

– estimating for each simplified PPE:

- the percentage of each characteristic vegetation class;

- the distribution of each characteristic vegetation class within each simplified PPE and determine the overall dominant vegetation based on a defined decision rule.

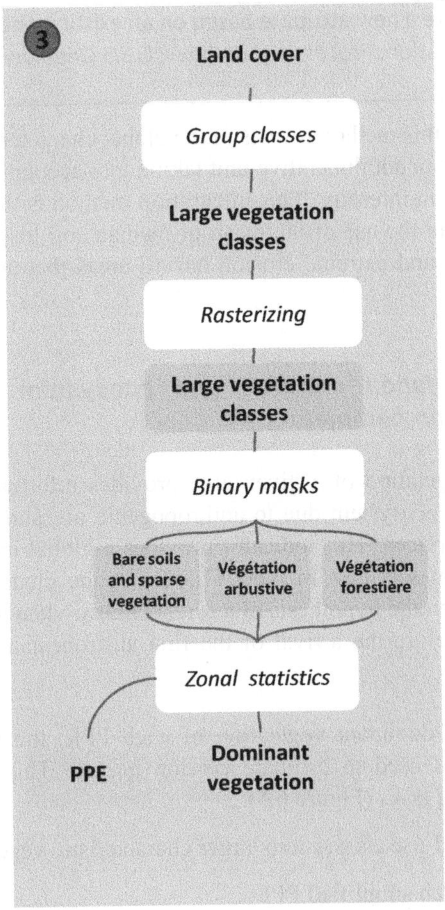

Figure 6.5. *Dominant vegetation determination of each simplified PPE*

6.2.3.1. *Determination of large vegetation classes*

The New Caledonia land cover database produced by the *Direction des Technologies et des Services de l'Information* of the Government of New Caledonia (DTSI), adapted to its context, consists of 19 classes grouped into four categories. This classification results from a semiautomatic procedure of satellite imagery analysis. Its validity is therefore not absolute. The classification was validated with a kappa coefficient of 75.5% [DIR 08].

The four initial categories do not correspond to large classes of vegetation. The 19 detailed land-use classes are thus aggregated into three classes including (Table 6.3).

Land use database	Large vegetation class
Forest on ultrabasic soils	Forest
Forest on volcanic and sedimentary soils	
Shrublands on volcanic and sedimentary soils	Shrublands and herbaceous vegetation
Savanah	
Dense maquis	
Herbaceous maquis	
Sparse vegetation on volcanic and sedimentary soils	Baresoils and sparse vegetation
Volcanic or sedimentary bare soils	
Sparse vegetation on basic soils	
Ultrabasic bare soils	
Urban areas	*Not reclassed*
Topographic and cloud Shadows	
Water	
Ocean	
Sparse mangroves	
Dense mangroves	
Clouds	
Salted bare soils	

Table 6.3. *Aggregation of land use classes in large vegetation classes in New Caledonia [AND 16]*

QGis function to create a new attribute based on an existing one:
- New attribute calculation: *Processing toolbox > QGIS Geoalgorithms > Vector table tools > Field calculator...*

6.2.3.2. *Determination of the dominant vegetation within PPE*

The determination of the dominant vegetation in New Caledonia's PPE is defined by the dominance of one or a set of large classes of vegetation [AND 16].

Three types of dominant vegetation are defined:

– dominant vegetation 1: landscape dominated by tree and shrub formations corresponding to low degraded to degraded forests;

– dominant vegetation 2: mosaic of tree, shrub and herbaceous formations with the presence of bare soils and predominance of shrub and herbaceous formations corresponding to degraded to very degraded forests;

– dominant vegetation 3: landscape with the large presence of bare soil where the vegetation is sparse and consists of herbaceous formations, with some shrub and tree formations from very degraded to extremely degraded forests.

From the large vegetation map, the respective percentage of each class is calculated within each PPE.

QGIS function to calculate zonal statistics:
- Zonal statistics: *Raster > Zonal statistics > Zonal statistics ...*

Then, the dominant vegetation of each PPE is determined using a decision rule (Table 6.4) taking into account the percentages of areas of each large vegetation class in each PPE.

Percent of land covered in the PPE	Dominant vegetation type
• Forest >59% • Or Forest cover ≤59% and Shrublands and herbaceous vegetation >33% and Shrublands and herbaceous vegetation ≤68% and Bare soils and sparse vegetation ≤11%	**Dominant vegetation 1**: landscape dominated by tree and shrub formations
• Shrublands and herbaceous vegetation >68% • Or Forest ≤59% and Bare soils and sparse vegetation ≤36% and Bare soils and sparse vegetation >11% • Or Forest ≤59% and Shrublands and herbaceous vegetation >33%	**Dominant vegetation 2**: mosaic of tree, shrub and herbaceous formations with the presence of bare soils and predominance of shrub and herbaceous formations
• Forest ≤59% and Bare soils and sparse vegetation >36% and Forest cover >20% • Or forest cover ≤20% and Shrublands and herbaceous vegetation ≤33%	**Dominant vegetation 3**: landscape with the large presence of bare soil where the vegetation is sparse and consists of herbaceous formations, with some shrub and tree formations from degraded forests

Table 6.4. *Dominant vegetation determination of New Caledonia's PPE [AND 16]*

QGIS function to create a new attribute based on an existing one:
- New attribute calculation: *Processing toolbox > QGIS Geoalgorithms > Vector table tools > Field calculator...*

6.2.4. *Resilience criterion: forest fragmentation parameter*

This is the fragmentation in itself, qualified independently of the loss of habitat (qualified by the dominant vegetation). The fragmentation characterizes the forest mosaic and indicates the loss of fire resilience. This parameter also provides information on ecosystem degradation due to anthropogenic pressures, and services or associated loss of services.

Forest fragmentation was estimated for each PPE using the formulas proposed by [HUR 02, MOU 12, RIT 00]. Several indices exist to estimate forest fragmentation. In the scientific literature, experts often use a large number of indices to characterize the forest fragmentation.

Two global fragmentation indices were chosen: the shape index and the diversity index. The easily reproducible procedures to estimate these indices at the territorial scale in exploiting only the forest coverage data from land-use data explain this choice [AND 16].

The fragmentation estimates for each PPE involve three steps (Figure 6.6):

– the calculation of indices is based on the shape, the number and the distribution of forests within each PPE. The initial step consists of the intersection between the PPE database (section 6.2.1) and the large forest class determined in section 6.2.2.1;

– from the geometric attributes of each forest polygon, the two fragmentation indices (shape and diversity) are calculated using zonal statistics in spreadsheet format. They are then assigned to the corresponding PPE by an attribute table join;

– forest fragmentation within the PPE is estimated according to a defined decision rule.

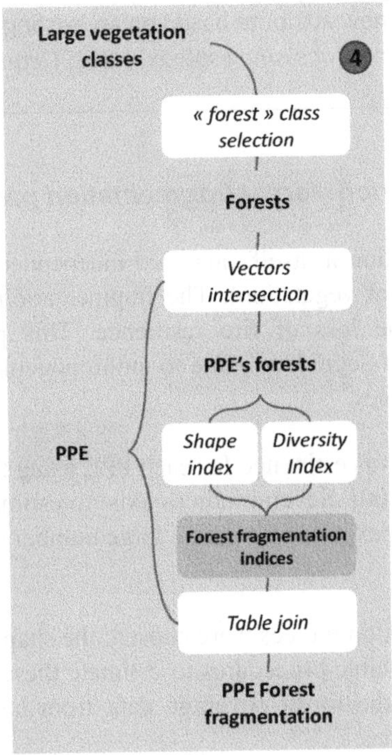

Figure 6.6. *Forest fragmentation index determination method*

6.2.4.1. *Forest class intersection with the PPE database*

First, the forest class is extracted from the classification of the large vegetation types. This class is then intersected with the PPE database in order to keep only the forests in the PPE polygons.

> **QGIS function to intersect the forest class with PPE:**
> - Select entity with an expression on attributes
> - Save the selection as a new vector layer...
> - Vector intersection: *Processing toolbox > QGIS Geoalgorithms > Vector overlay tools > Intersection*

6.2.4.2. *Diversity index*

The diversity of forest patches, denoted $H_j(a)$, was calculated using the Shannon index formula [BOG 05], equation [6.1], where *ln* is the natural logarithm and a_{ij} is the area of the *i*th patch of the forest *j*. Diversity indices founded on Shannon's index are based on the principle that the greater the diversity is, the greater the disorganization (entropy) is. This type of index is usually used to measure the relative abundance of ecosystem [FAO 99]:

$$H_j(a) = \sum_{i=1}^{n_j} - \left(\frac{a_{ij}}{\sum_{i=1}^{n_j} a_{ij}} ln \frac{a_{ij}}{\sum_{i=1}^{n_j} a_{ij}} \right). \qquad [6.1]$$

This index measures the relative diversity of forest patches. $H_j(a)$ value will depend on the number of patches (n_j), their relative proportion and their heterogeneity. The index is equal to 0 when the class consists of only one patch and its value will increase with the number of patches and with the equitability between the patches' areas of the class [MCG 95] (Figure 6.7):

1) a low diversity index value (close to 0) indicates that forest patches are very few and homogeneous;

2) a diversity index value close to 1 indicates a slight diversification of forest patch shape types, a low number of patches and of homogeneous size;

3) a diversity index value close to 2 indicates a greater heterogeneity of the forest patches, a greater number of patches and of more heterogeneous sizes;

4) a diversity index value greater than 2 corresponds to a large number of forest patches heterogeneous in type, shape and size.

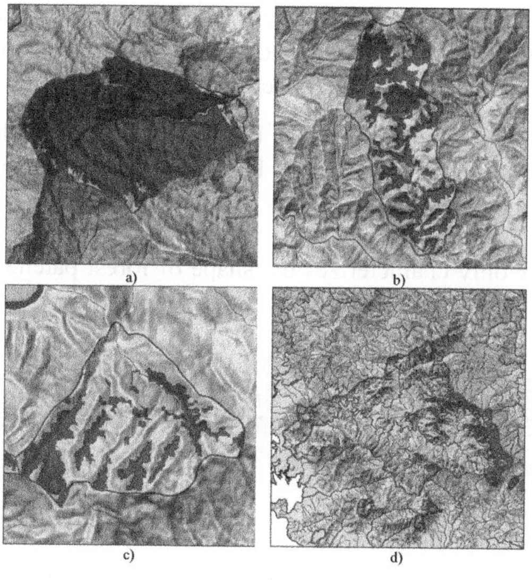

Figure 6.7. *Examples of forest diversity indices in New Caledonia. a) Diversity index = 0.02. b) Diversity index = 0.90. c) Diversity index = 1.95. d) Diversity index = 3.56. For a color version of the table, see www.iste.co.uk/baghdadi/qgis3.zip*

QGIS function to create a new attribute based on an existing one:
- New attribute calculation: *Processing toolbox > QGIS Geoalgorithms > Vector table tools > Field calculator...*

QGIs function to calculate group statistics for selected attributes:
- group statistics: *Vector > Group Stats > GroupStats*

6.2.4.3. *Shape index*

The shape index of the forest j (IF_j) is determined using the following equation:

$$IF_j = \frac{P_{ij}^2}{\sum_{i=1}^{n_j} a_{ij}}, \qquad [6.2]$$

where P_{ij} is total perimeter of the class j and a_{ij} is the area of the ith patch of the forest j.

The shape index is based on the ratio of the perimeter on the area. The more the patches have elongated or irregular shapes, the higher the IF_j value will be and this value will decrease as the shapes become circular [BOG 00] (Figure 6.8):

– low shape index values (0 to 50) characterize compact objects approaching a circular shape. a perfect circular object has a shape index value equal to 4;

– shape index values close to 100 characterize objects with progressi0vely complex shapes such as, for example, a compact forest patch in upper watershed with a descending tongue running along a creek at the bottom of the talweg;

– high shape index values (above 1,000) indicate complex and ramified (dendritic) forms. The higher the index is beyond this value, the more complex the patterns of forest patches are.

The shape index only characterizes the shape of forest patches. It does not give information on the number of patches, nor on their homogeneity or heterogeneity.

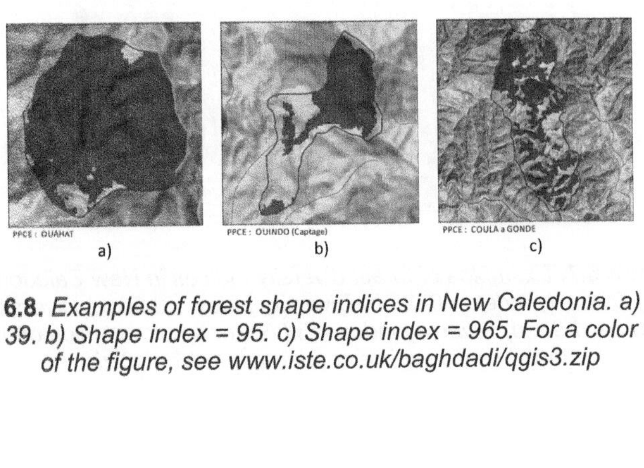

Figure 6.8. *Examples of forest shape indices in New Caledonia. a) Shape index = 39. b) Shape index = 95. c) Shape index = 965. For a color version of the figure, see www.iste.co.uk/baghdadi/qgis3.zip*

QGIS function to create a new attribute based on an existing one:
- New attribute calculation: *Processing toolbox > QGIS Geoalgorithms > Vector table tools > Field calculator...*

QGIS function to calculate group statistics for selected attributes:
- group statistics: *Vector > Group Stats > GroupStats*

6.2.4.4. *PPE Forest fragmentation*

The determination of forest fragmentation intensity within each PPE is performed in two steps from the two previous indices:

– the first step consists of defining threshold values for these two indices corresponding to classes of forms and fragmentation of forest patches;

– the second step consists of the cross-analysis of the form classes and the forest patches fragmentation according to a defined decision tree in order to estimate the global forest fragmentation of each PPE.

The threshold values used for each of the two indices are presented in Table 6.5.

Forest patches shape	Shape index thresholds	Forest patches diversity	Diversity index thresholds
Compact to complex patch	[0; 400[	Few homogeneous patches	[0; 0.6[
Highly complex patch	[400; 2,250]	More patches associated with more heterogeneous shape	[0.6; 1.9]
Ramified and highly complex patch	[2,250; +∞[	A lot of highly heterogeneous patches	[1.9; +∞[

Table 6.5. *Threshold values used to define forest fragmentation classes of New Caledonia's PPE [AND16]*

The determination of the PPE's forest fragmentation is carried out according to a decision tree from three classes of forest shape and three classes of forest diversity (Figure 6.9). Forest fragmentation is then qualified logically.

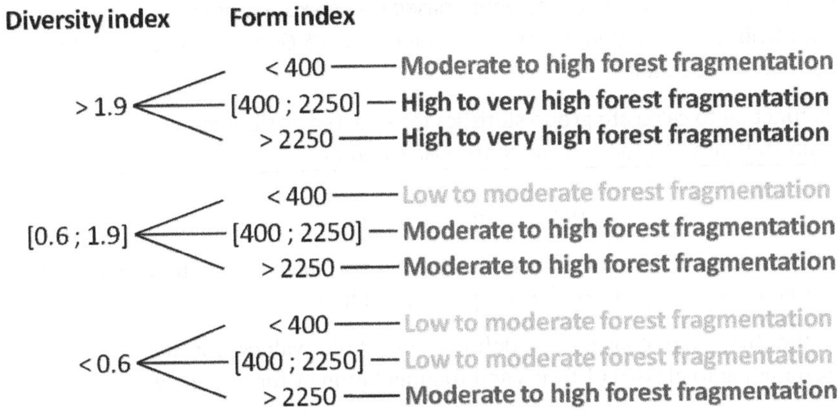

Figure 6.9. *PPE's forest fragmentation decision tree. For a color version of the figure, see www.iste.co.uk/baghdadi/qgis3.zip*

QGIS function to create a new attribute based on an existing one:
- New attribute calculation: *Processing toolbox > QGIS Geoalgorithms > Vector table tools > Field calculator...*

6.2.5. *Assessment of the forest function in the PPE*

The decision tree (Figure 6.10) that allows us to characterize the function state of a forest is based on the concatenation of the three parameters, “erosion hazard”, “dominant vegetation” and “forest fragmentation” using the Boolean operator “AND”. For each possible combination, a function state is assigned logically. For example, the concatenation of the criterion “Strong to extreme erosion hazard” AND “Dominant vegetation 3” AND “High to very high forest fragmentation” correspond to the assignment of a “highly deteriorated” forest function state.

The ecosystem functionality analysis is qualitative and is based on a service efficiency logic (for example a PPE characterized by high erosion hazard with forest as dominant vegetation will be assigned as “highly deteriorated”: the ecosystem's soil protection service is failing). Nevertheless, a greater emphasis was placed on the “dominant vegetation” parameter due to a loss of global services (water regulation, soil stabilization, fire resilience, supporting services) compared to the erosion hazard parameter.

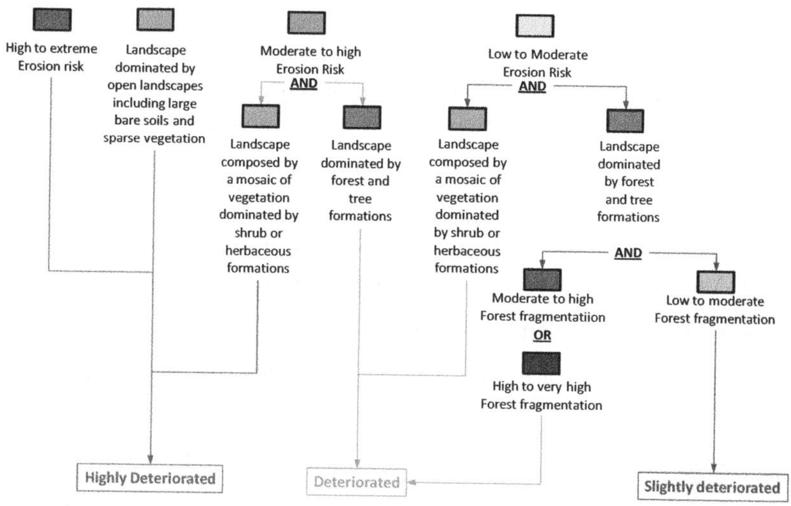

Figure 6.10. *The decision tree used to assess forest function state of the PPE (dominant vegetation 1: landscape dominated by tree and shrub formations; Dominant vegetation 2: mosaic of tree, shrub and herbaceous formations with the presence of bare soils and predominance of shrub and herbaceous formations; Dominant vegetation 3: landscape with the large presence of bare soil where the vegetation is sparse and consists of herbaceous formations, with some shrub and tree formations from degraded forests). For a color version of the figure, see www.iste.co.uk/baghdadi/qgis3.zip*

Forest fragmentation gives a greater degree of sharpness to the estimate of the dominant vegetation and makes it possible to distinguish between "Slightly deteriorated" and "deteriorated" forest function states, in the case that erosion hazard is qualified as low to moderate where the landscape is dominated by tree and shrub formations.

Thus, the forest function state of PPEs having only:

– the landscape dominated by tree and shrub formations (dominant vegetation type 1);

– low to moderate erosion hazard;

– low to moderate forest fragmentation;

will be assigned as slightly deteriorated.

QGIS function to create a new attribute based on an existing one:
- New attribute calculation: *Processing toolbox > QGIS Geoalgorithms > Vector table tools > Field calculator...*

6.2.6. *Limits*

The three parameters are not completely independent of one another. The approach described here is qualitative and weighted according to these parameters.

Indeed, the erosion hazard includes the vegetation cover (derived from land cover), in addition to rainfall and runoff variables, slope and slope length, and soil erodibility. The vegetation cover is directly used to determine the dominant vegetation. And each parameter exploited also has intrinsic limitations:

– soil loss estimates are derived from modeling;

– the forest land cover class does not characterize the degradation degree of the understory vegetation;

– the DTSI land cover databases have a certain margin of inaccuracy (Kappa coefficient of 75.5%).

6.3. Forest function assessment implementation

This section presents a practical implementation of the RUSLE model on the Dumbéa area in New Caledonia (South-West Pacific). This area is characterized by a mountain chain reaching 1,200 m high and subject to heavy erosive processes due to runoff. The mountain range is bordered by a *piedmont glacis* followed by a coastal plain on the West Great Lagoon of the Grande Terre of New Caledonia. Since the early 2000s, this area has experienced forest degradation due to bush fires in particular and leading to direct impacts on the quality of water resources [AND 16].

6.3.1. *Software and data*

6.3.1.1. *Software*

Raster processing and vector analysis are performed in this implementation with the OSGeo4W software suite that includes QGIS[1] version 2.18 with the GRASS extension [LAC 17].

The OSGeo4W setup program can be downloaded at the following link: https://www.qgis.org/fr/site/forusers/download.html

1 QGIS is a user-friendly GIS software distributed under the GNU public licence. It is an official project of the Open Source Geospatial foundation (OSGeo) [QGI 17].

Some processing used here requires the installation of three additional QGIS extensions:

– *Group Stats* Version 2.0.30: extension developed by Rajmund Szostok allowing group analysis and statistics with vector data;

– *Geometry check* Version 1.0: used to perform geometry checks and corrections to vector data;

– *Zonal Statistics* Version 0.1: allowing zonal statistics of a raster layer inside vector polygons (count, measures of central tendency).

Approach	Practical work in QGIS
1. Plugins repository	In **QGIS**. In the **menu bar**: • Go to: **Plugins** > **Manage and install plugins**
2. Search and install extensions	• In the search bar of the module:  ○ Search the Group Stat plugin ○ The plugin list automatically filters the plugins. 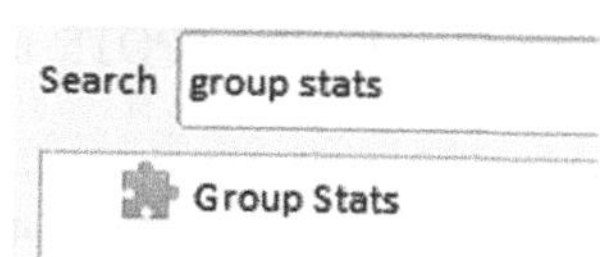 • Select the plugin and click on "Install plugin" • Installation may take a little while • When the plugin is correctly installed, it is checked in the available plugin list 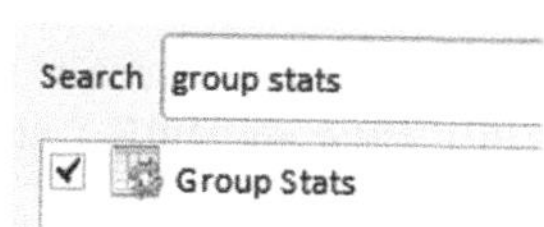 • Proceed with the same procedure to install the Geometry Check and Zonal Statistics extensions. • Close the plugin manager window.

Table 6.6. *Install QGIS extensions*

6.3.1.2. *Data: downloading of land cover*

The Water Catchment Perimeters vector file is provided for this implementation (DAVAR_PPE_NC_SELECTION.shp). This is a selection made from the Water Catchment Perimeters database by the Department of Veterinary and Food Affairs of New Caledonia (DAVAR) in charge of the perimeters inventory and the database maintenance for mandatory purposes in New Caledonia.

Soil loss estimates from the RUSLE equation obtained in the chapter [AND 17] can be directly used for this implementation as they cover the same area. However, soil loss estimates are also provided for this implementation in TIF format (pertes_en_sol_A.tif).

The land cover database used in this implementation can be downloaded from the New Caledonia geographical web server: http://georep.nc/. Data are distributed under a creative commons license (the use conditions for each dataset are described in their metadata).

The land cover classification is the result of a semi-automatic processing procedure using SPOT satellite imagery for the period 2006–2008. The 19 classes are grouped into four categories: water and mangroves, urban area, vegetation on volcanic and sedimentary rocks, and vegetation on ultrabasic rocks [DIR 08]. The land cover database is delivered in shapefile format within the reference coordinate system RGNC91-93.

File name: occupation_du_sol_2008_SPOT5_approche_objet_shapefile.zip

Link: http://www.geoportal.gouv.nc/geoportal/catalog/search/resource/details.page?uuid=%7B44E58684-49BA-4F60-AFB7-BB24464FF4B4%7D

File size: 71.6 Mb.

6.3.2. *Step 1: PPE polygons creation*

The water catchment perimeters vector file includes three types of water protection perimeters (PPI, PPR and PPE_S) digitalized according on the spaghetti model.

The three polygons representing the PPI, PPR and PPE_S perimeters around the same water catchment area are grouped and cleaned to obtain a clean simplified layer of PPE polygons.

Approach	Practical work in QGIS
1. Open the water catchment perimeters vector file	In **QGIS:** • Open the vector file : o DAVAR_PPE_NC_SELECTION.shp
2. Check attributes table	In **QGIS:** • Open the attribute table of the vector file by selecting the layer, right-click and click on Open Attribute Table. **Open Attribute Table** • The layer contains an attribute named ID_PPE which is a unique identifier per water catchment reported for each PPI, PPR and PPE_S.
3. Merge shape files according to an attribute	In the **menu bar**: • Click on: **Processing > Toolbox** The QGIS Toolbox opens. In the **QGIS Geoalgorithms**: • Go to: **Vector geometry tools > Dissolve** • Select the DAVAR_PPE_NC_SELECTION.shp layer. • Uncheck the option "Dissolve all (do not use field)" • Select the field ID_PPE as Unique ID field. • Save the new vector file as DAVAR_PPE_NC_SELECTION_DISSOVLE.shp The polygons are merged by ID_PPE 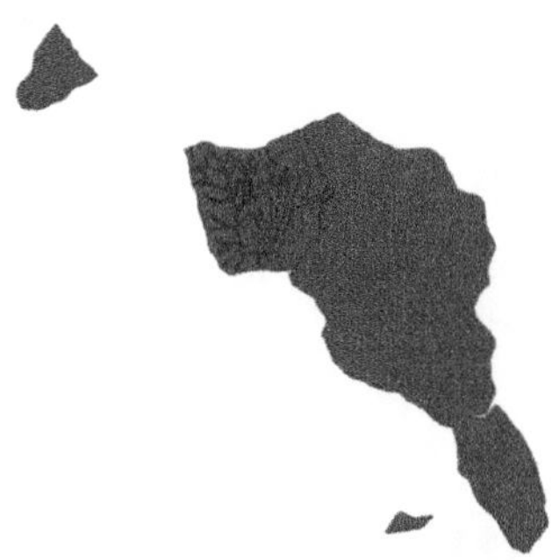

Table 6.7. *Merging shapes according to an attribute*

Some perimeters have defects (overlap, geometric errors) that need to be corrected. The correction is made in two steps:

– using the preinstalled Geometry Check plugin;

– using the cleaning tool of the GRASS toolbox.

The Geometry Check is one of the main plugins of QGIS. It allows the validity of a geometry of a given layer to be checked and repaired. The geometry check plugin can find the following errors (Figure 6.11):

a) Self-intersections: a polygon with a self-intersection;

b) Duplicate nodes: two duplicates nodes in a segment;

c) Holes: hole in a polygon;

d) Segment length: a segment length lower than a threshold;

e) Minimum angle: two segments with an angle lower than a threshold;

f) Minimum area: polygon area lower than a threshold;

g) Silver polygon: this error comes from a very small polygon (with small area) with a large perimeter;

h) Duplicate features;

i) Features within features;

j) Overlaps: polygon overlapping;

k) Gaps: gaps between polygons.

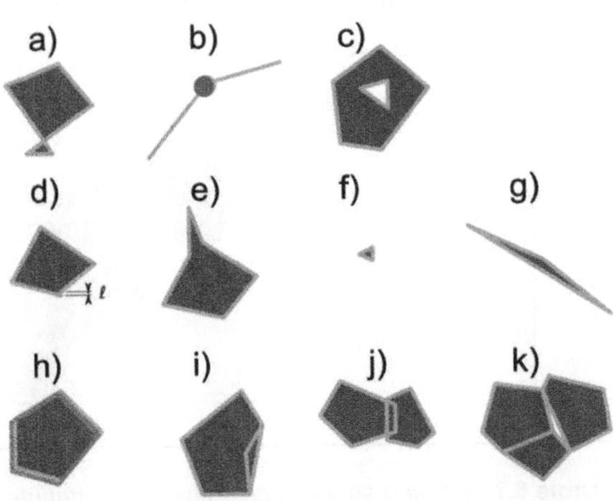

Figure 6.11. *The Different checks supported by the plugin*

Approach	Practical work in QGIS
1. Open the Geometry check plugin	In the **menu bar**: • Go to: **Vector > Geometry tools > Check geometries** Check Geometries
2. Configure the geometry check window (QGIS < 2.18)	In the window of the geometry checker: • Select DAVAR_PPE_NC_SELECTION_DISSOVLE.shp as input vector layer; • Verify that the option "Only selected features" is unchecked; • In the section "**Geometry validity**" select the following options: o Self intersections o Duplicate nodes o Polygon with less than three nodes • In the section "**Allowed geometry types**", select the option "Polygon" in order to allow this type of geometry only; • In the section "**Geometry properties**", select the option "Polygons and multipolygons may not contain any holes" in order to avoid any discontinuity or rings in the polygons of the layer; • In the section "**Geometry conditions**", select the options with their corresponding validation parameters as follow: o Minimum segment length (map unit) with the value 1; o Minimum angle between segments (deg) with the value 0.1; o Minimal polygon are (map unit sqr.) with the value 1; o No sliver polygons (Maximum thinness set to 10 and Max. area (map unit sqr.) unchecked); • In the section "**Topology checks**", select the options and their corresponding parameters as follow: o Check for duplicates; o Check for features within other features; o Check for overlaps smaller than (map units sqr.) with the default value 0.00; o Check for gaps smaller than (map units sqr) with the default value 0.00; • Leave **Tolerance** to the default value 1E-8; • In the section "**Output vector layer**", select "Modify input layer" to perform interactive correction with the tool.

	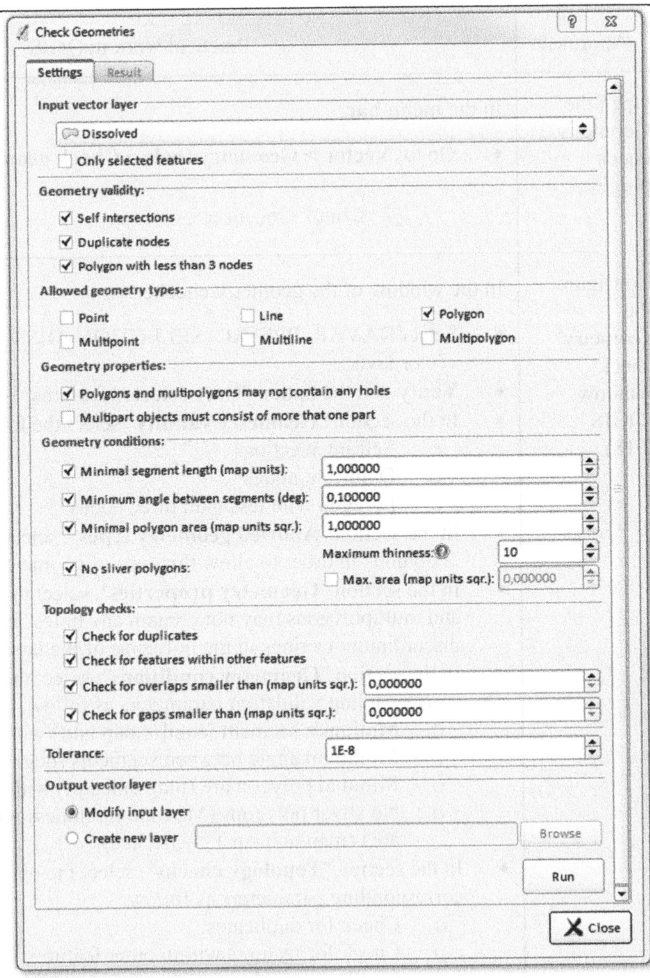 • Click "Run" to run the analysis.
3. Understand the results of the interactive geometry checker	The results appear in the **"Result"** tab and are visible in the canvas as a pointer to the invalid geometries. A table lists the **"geometry check result"** with one error per line: the first column is an identifier of the object, the second is the reason for the error, then the coordinates of the error, a value (in the case when the error is estimated with respect to a given threshold, such as a minimum allowed angle), and finally the resolution column that indicates the resolution of the error. At the bottom of this table, you can export the error to a shapefile. To the right, you have the total number of errors and errors corrected.

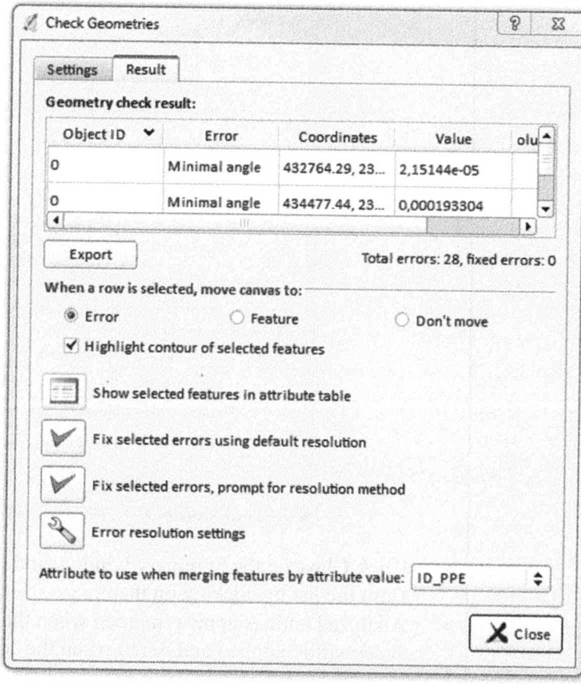

- At the bottom right of the interactive geometric error correction window, select the field ID_PPE to be used when merging entities by attribute value;
- Click  to "**Fix selected errors using default resolution**";
- Click "**Yes**" in the pop-up window asking if you want to correct the total number of errors identified;
- A **Summary** window opens including from the top to the bottom:
    - Fixed errors and their number;
    - Unfixed errors and their number;
    - new errors found and their number;
    - Obsolete errors due to the corrections applied;

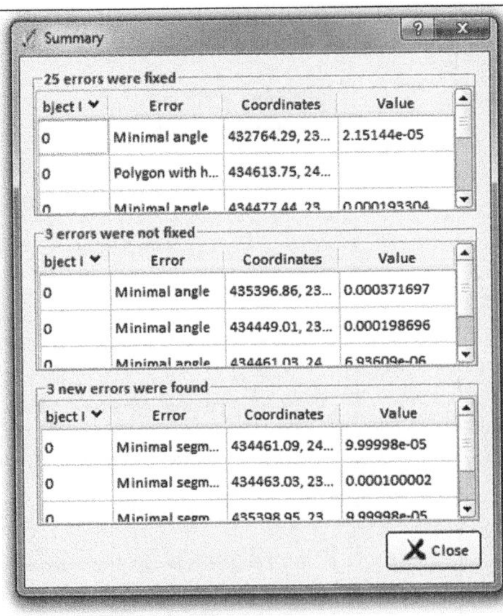

- Click **Close** on the Summary window and select the new entities from the list by clicking on them;
- Analyzed entities appear in green when the correction has been successfully applied and in red when the correction has failed;

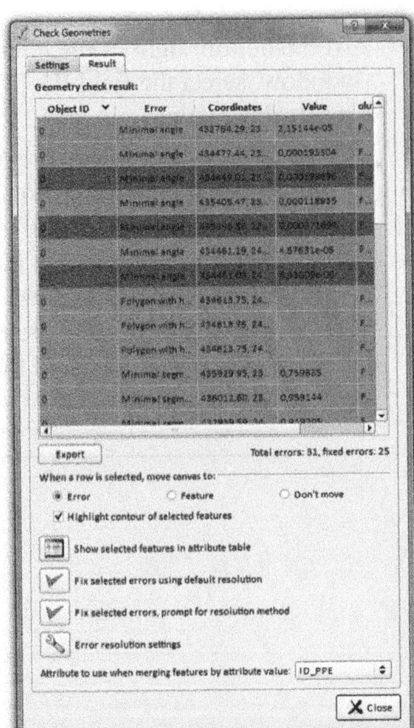

- Restart the automatic correction on the new detected and selected entities until the summary window no longer indicates new errors;

<table>
<tr><td></td><td>• Select the entities for which auto-correction has failed and click  to fix the errors in choosing the type of correction method to use, and test alternative methods of geometric corrections;
• The geometric check can be started several times since the corrections result in modifications of the vector layer and the latter may not be in agreement with the geometric rules previously specified.</td></tr>
</table>

Table 6.8. *Geometry check. For a color version of the table, see www.iste.co.uk/baghdadi/qgis3.zip*

To ensure a complete correction of the polygon layer, additional processing using the cleaning tool of the GRASS toolbox allows residual errors to be corrected that the geometry check plugin could not process. The GRASS cleaning function incorporates a specific function to remove small polygons; the analysis is performed to remove polygons smaller than 1 ha.

Approach	Practical work in QGIS
1. Use the GRASS v.clean function	In the **Processing Toolbox**: • Click on: **GRASS Commands > Vector > v.clean** • Select the layer to clean DAVAR_PPE_NC_SELECTION_DISSOLVE.shp • From the dropdown list "Cleaning tool", you can choose between several cleaning and geometry correction options (angles correction, snapping, remove small polygons, etc.). Select the **rmarea** option that removes small polygons; • Set 10,000 as the **Threshold** value. rmarea function will remove all the polygons smaller than 1ha. They will be merged to the neighboring polygon sharing the longest border with the polygon removed. • Leave default value for the option "GRASS region extent" to process all the data in input; • Save the cleaned layer as DAVAR_PPE_NC_SELECTION_SIMPLIFIE.shp; • Leave the default value for the errors output file (Save to temporary file) and uncheck the option "Open output file after running algorithm".

<table>
<tr>
<td></td>
<td>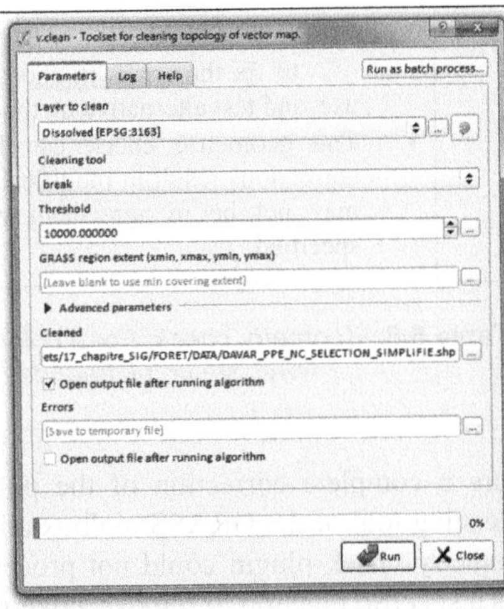
</td>
</tr>
<tr>
<td>2. Visual check</td>
<td>Open the newly created cleaned file and verify that the polygons have no apparent geometric defects.

Original PPE polygons:

Merged and cleaned PPE polygons:
</td>
</tr>
</table>

Table 6.9. *Remove small polygons using GRASS. For a color version of the table, see www.iste.co.uk/baghdadi/qgis3.zip*

6.3.3. *Step 2: erosion hazard parameter determination*

The "erosion hazard" parameter determination consists of analyzing the distribution of erosion classes in each PPE perimeter.

6.3.3.1. *Creation of erosion hazard mask from RUSLE estimates*

Erosion hazard class determination consists of the creation of four binary masks, each corresponding to a specific interval of soil loss estimates from the RUSLE equation. The threshold values are those defined in section 6.2.1.1.

Approach	Practical work in QGIS
1. Open the soil loss estimates from the RUSLE equation	In **QGIS**: • Open the RUSLE soil loss estimates raster layer: pertes_en_sol_A.tif 
2. Create the binary mask "Low erosion hazard"	In the **menu bar**: • Click on **Raster > Raster calculator…** In the **Raster calculator**: • enter the following expression: "pertes_en_sol_A@1"< 2

	Save the output layer as: alea_erosif_faible.tif The result is a binary raster (0 and 1 values) where the value 1 corresponds to soil loss estimates smaller than 2 t/ha/year. 
3. Create the binary masks "Moderate erosion hazard", "High erosion hazard", "Extreme erosion hazard"	In the **Raster calculator**: • enter the following expression: "pertes_en_sol_A@1">=2 AND "pertes_en_sol_A@1"<23.3 Save the output layer as: alea_erosif_modere.tif In the **Raster calculator**: • enter the following expression: "pertes_en_sol_A@1">=23.3 AND "pertes_en_sol_A@1"<102.1 Save the output layer as: alea_erosif_forte.tif In the **Raster calculator**: • enter the following expression: "pertes_en_sol_A@1">= 102.1 Save the output layer as: alea_erosif_extreme.tif At the end of this step, four raster files are created corresponding to each class of erosion hazard intensity based on soil loss estimates.

Table 6.10. *Creation of the mask erosion hazard*

6.3.3.2. *Zonal statistics of erosion hazard mask in PPE polygons*

The estimates of the percentage of each erosion hazard class in each PPE polygon is carried out by zonal statistics between the binary masks and the PPE vector layer. Indeed, the mean of the values of the binary mask pixels, whose possible values are 0 (absence) and 1 (presence), is equivalent to the calculation of the percentage of the masks' values for each polygon.

Approach	Practical work in QGIS
1. Open the erosion hazard binary masks	In **QGIS**: • Open the binary masks previously created: alea_erosif_faible.tif alea_erosif_modere.tif alea_erosif_fort.tif alea_erosif_extreme.tif • Open the PPE vector layer: DAVAR_PPE_NC_SELECTION_SIMPLIFIE.shp 
2. Zonal statistics calculation	In the **menu bar**: • Go to **Raster > Zonal statistics > Zonal statistics** • Choose the raster layer alea_erosif_faible.tif • Verify that the selected band is band 1 • Choose DAVAR_PPE_NC_SELECTION_SIMPLIFIE as the polygon layer containing the zones • Enter "EFA" as Output column prefix. The calculation of zonal statistics creates new fields in the vector attribute table. The choice of the prefix must be clear enough to

<table>
<tr>
<td></td>
<td>
quickly know the content of the field and be unique to avoid overwriting an existing field.

- Select "Mean" as "Statistics to calculate":

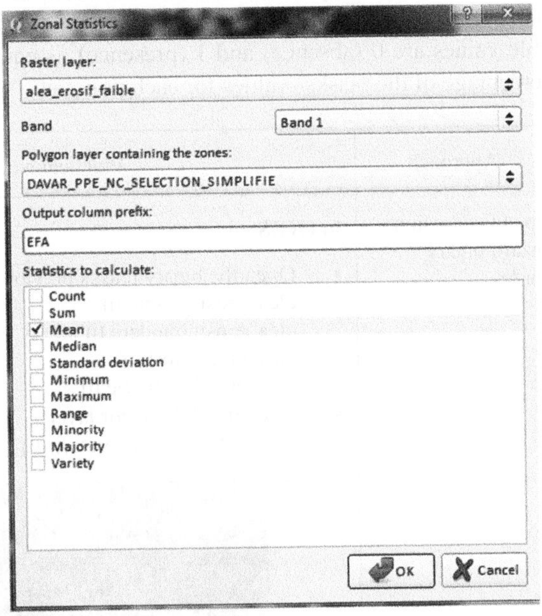

Click OK.
</td>
</tr>
<tr>
<td>3. Check the results</td>
<td>
Check the results of the Zonal statistics calculation by selecting the layer DAVAR_PPE_NC_SELECTION_SIMPLIFIE.shp, right-click and Open the attribute table

A new field was created named "EFAmean"
</td>
</tr>
<tr>
<td>4. Calculate the zonal statistics for each erosion hazard mask</td>
<td>
Repeat the operation with three other erosion hazard masks:

alea_erosif_modere.tif
alea_erosif_fort.tif
alea_erosif_extreme.tif

by naming the Output column prefix, respectively, as follows:

EFM
EFF
EFE

At the end of this step, the vector layer file has four additional fields EFAmean, EFMmean, EFFmean and EFEmean.
</td>
</tr>
</table>

EFAmean	EFMmean	EFFmean	EFEmean
0.2229755486...	0.0812775690...	0.0258127756...	0.6699341066...
0.3493592878...	0.5837816687...	0.0043040203...	0.0625550229...
0.3644859813...	0.4314641744...	0.0031152647...	0.2009345794...
0.4034821605...	0.3157969116...	0.0082060655...	0.2725148623...
0.4185260311	0.2816091954	0.0144241604	0.2854406130

In order to check the validity of the calculations, you can sum these four new fields in a temporary field. The sum must be equal to 1 for each line.

Table 6.11. *Calculation of zonal statistics of erosion hazard*

6.3.3.3. *PPE erosion hazard determination*

This step consists of applying the established decision rule to classify each PPE according to the percentage of each erosion hazard class. This processing is performed by calculating a new attribute from the four fields calculated in the previous step.

For the ease of processing in the next steps, the PPE erosion hazard is coded as a discontinuous integer whose values are:

– 1: low to moderate erosion hazard;

– 2: moderate to high erosion hazard;

– 3: high to extreme erosion hazard.

Approach	Practical work in QGIS
1. Update the attribute table	Open the attribute table of the layer DAVAR_PPE_NC_SELECTION_SIMPLIFIE.shp • Open the field calculator in clicking on In the **Field calculator**: • Select **Create a new field** • Enter the **Output field name**: A_EROSIO • Select the **Output field Type** as: Whole number (integer) • Specify the **Output field length** = 10 • Enter the following condition in the **Expression tab**:

	CASE WHEN (("EFEmean">0.5) OR ("EFEmean">0.18 AND "EFFmean">0.48) OR ("EFEmean">0.18 AND "EFFmean"+"EFMmean">0.29) OR ("EFEmean"+"EFFmean"+"EFMmean">"EFAmean" AND "EFEmean">"EFFmean"+"EFMmean") OR ("EFEmean"+"EFFmean"+"EFMmean">0.79 AND "EFMmean"<"EFFmean"+"EFEmean") OR ("EFEmean">0.18 AND "EFFmean">0 AND "EFFmean"+"EFMmean">0.29)) THEN 1 WHEN (("EFEmean">0.18 AND "EFFmean">0 AND "EFFmean"+"EFMmean"<0.29) OR ("EFEmean">0 AND "EFEmean"<=0.18 AND "EFAmean"<0.22) OR ("EFFmean">0.48 AND "EFEmean">0 AND "EFEmean"<=0.18) OR ("EFEmean"+"EFFmean"+"EFMmean">"EFAmean" AND "EFEmean"<"EFFmean"+"EFMmean")) THEN 2 WHEN (("EFAmean">0.22 AND "EFEmean"+"EFFmean"<0.2) OR ("EFEmean"+"EFFmean"+"EFMmean"<"EFAmean")) THEN 3 ELSE 0 END
	• Click **OK** 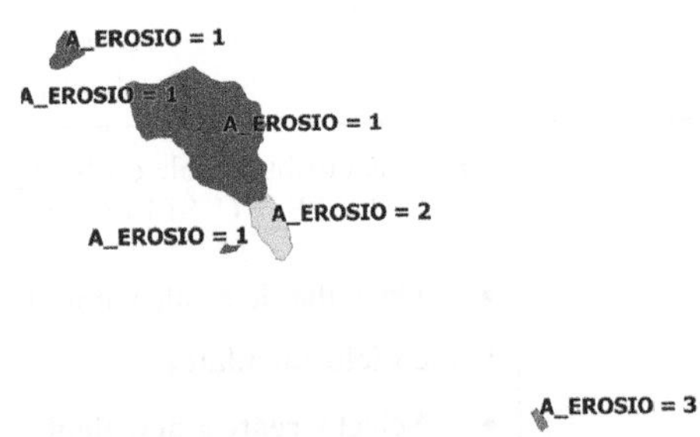 At the end of this step, the field A_EROSIO corresponds to the erosion hazard parameter and qualifies each PPE as 3: low to moderate erosion hazard, 2: moderate to high erosion hazard, 1: high to extreme erosion hazard.

Table 6.12. *Estimation of the erosion hazard for each PPE. For a color version of the table, see www.iste.co.uk/baghdadi/qgis3.zip*

6.3.4. *Step 3: dominant vegetation type parameter determination*

The dominant vegetation type parameter is derived from the land cover database by following the same method as for the erosion hazard parameter, i.e.:

– creating three raster masks corresponding to the large vegetation types;

– calculating the percentage of each class in the PPE polygons;

– computing a new field that classifies each PPE in its dominant vegetation type class by applying the decision rule defined in section 6.2.2.2.

6.3.4.1. *Land cover class aggregation in larger vegetation types*

The land cover database covers all the New Caledonia territory. In order to optimize the processing time, the first step is to clip the land cover database to our area of interest. We clip the land cover database according to the RUSLE erosion model layer area.

Approach	Practical work in QGIS

1. Verify the input files	In **QGIS**:
	• Open the vector layer:
	  o occupation_sol_2008_spot5_objet.shp
	• Open the raster layer:
	  o pertes_en_sol_A.tif

2. Clip the Land cover vector file	In the **menu bar**:
	• Go to: **Processing > Toolbox**
	The QGIS Processing toolbox opens.
	In the **QGIS geoalgorithms**:
	• Select : **GDAL/OGR > Geoprocessing > Clip vector by extent**
	• Select the input layer:
	  occupation_sol_2008_spot5_objet.shp
	• For the parameter **Clip extent**:
	  o Choose **use layer/canvas extent**
	  o Choose the extent of the layer: pertes_en_sol_A.tif
	• Leave the additional creation options blank
	  o Save the clipped file as:
	  occupation_sol_2008_spot5_objet_decoupe.shp

<table>
<tr><td></td><td>Processing may take a few minutes. A window may pop-up to indicate different SCRs.

The result is a shapefile of land cover clipped on the raster boundaries of the RUSLE erosion model. In the attribute table, the CLASS field contains accents. Therefore, we must create an identical column and remove them before checking the geometric validity.</td></tr>
</table>

Table 6.13. *Clipping of the land cover map according to the extent of the study area*

From the clipped land cover layer, a new field is created to group the land cover classes according to the larger vegetation type they belong to. It is more convenient to encode this new attribute as a real number.

Land use database	Large vegetation class	Grid code
Forest on ultrabasic soils	Forest	1
Forest on volcanic and sedimentary soils		
Shrublands on volcanic and sedimentary soils	Shrublands and herbaceous vegetation	2
Savanah		
Dense maquis		
Herbaceous maquis		
Sparse vegetation on volcanic and sedimentary soils	Baresoils and sparse vegetation	3
Volcanic or sedimentary bare soils		
Sparse vegetation on basic soils		
Ultrabasic bare soils		
Urban areas	*Not reclassed*	0
Topographic and cloud shadows		
Water		
Ocean		
Sparse mangroves		
Dense mangroves		
Clouds		
Salted bare soils		

Table 6.14. *Large vegetation classes grid codes and their corresponding land cover classes*

Approach	Practical work in QGIS
Update the attribute table	Open the attribute table of the layer occupation_sol_2008_spot5_objet_decoupe.shp • Click on to open the Field calculator In the **Field calculator**: • Select **Create a new field** • Enter the **Output field name**: GT_VEG • Select the **Output field type**: Decimal number(real) • Enter **Output field length** = 12 • And enter **Precision** = 6 (number of digit after the decimal point) • In the **Expression tab**, enter the following condition: CASE WHEN ("CLASSE"='Forêt sur substrat ultramafique' OR "CLASSE"='Forêt sur substrat volcano-sédimentaire') THEN 1 WHEN ("CLASSE"='Maquis dense paraforestier' OR "CLASSE"='Végétation arbustive sur substrat volcano-sédimentaire' OR "CLASSE"='Savane' OR "CLASSE"='Maquis ligno-herbacé') THEN 2 WHEN ("CLASSE"='Sol nu sur substrat ultramafique' OR "CLASSE"='Sol nu sur substrat volcano-sédimentaire' OR "CLASSE"='Végétation éparse sur substrat ultramafique' OR "CLASSE"='Végétation éparse sur substrat volcano-sédimentaire') THEN 3 Else 0 End

	• Click **OK** 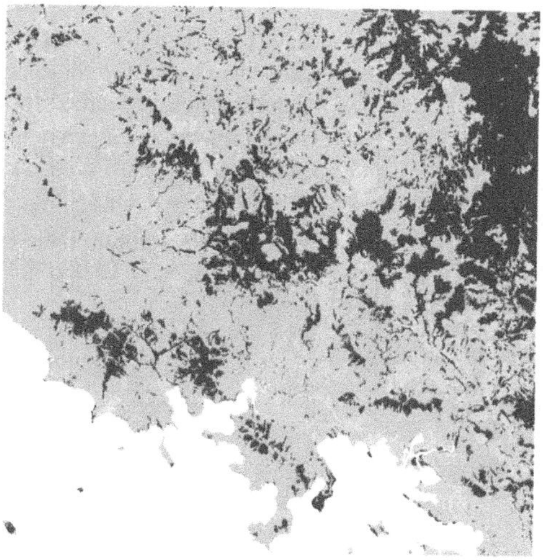 The clipped land cover layer has an new field, the values of which vary between 0 and 3: • 0: No vegetation; • 1: Forest; • 2: Shrublands and herbaceous vegetation; • 3: Bare soils and sparse vegetation.

Table 6.15. *Update the attribute table of the land cover map. For a color version of the table, see www.iste.co.uk/baghdadi/qgis3.zip*

This land cover layer is then rasterized according to the newly created field.

Approach	Practical work in QGIS
1. Verify the input	In **QGIS**: • Open the file: occupation_sol_2008_spot5_objet_decoupe.shp • Open the attribute table (select the file, right-click > **open the attribute table**) and verify that the field GT_VEG contains real number ranging from 0 to 3.
2. Rasterize a vector file	In the **menu bar**: • Go to: **Raster > Conversion > Rasterize (vector to raster)...**

- Select the input shapefile:

  occupation_sol_2008_spot5_objet_decoupe.shp
- Select the field GT_VEG as **Attribute field**
- Save the Output file for rasterized vectors as grands_types_vegetation.tif, and click OK
- A pop-up window opens indicating that the output file does not exist. You must set the file size or the resolution to be able to create it. click OK.
- Select the Resolution radio button expressed in map units per pixel and specify 10.000 in horizontal and vertical

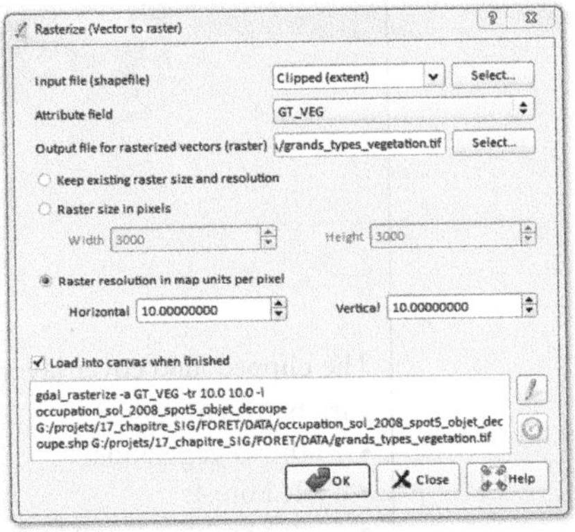

- Click OK
- Check the minimum and maximum values of the created raster that must be, respectively, 0 and 3 in the layer properties window, metadata tab .

The result is a raster file with values ranging from 0 to 3.

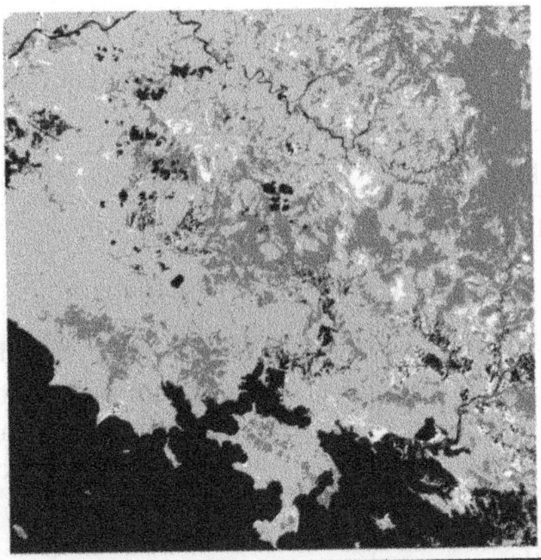

Table 6.16. *Rasterizing of the land cover vector file*

6.3.4.2. *Large vegetation types' masks*

From the large vegetation types raster file, three binary masks (0: absence; 1: presence) are created for each large vegetation type.

<table>
<tr><th>Approach</th><th>Practical work in QGIS</th></tr>
<tr><td>1. Create the forest binary mask</td><td>Open the file grands_types_vegetation.tif previously created,
In the menu bar:
• Go to Raster > Raster calculator…
In the Raster calculator:
• Enter the following expression:
"grands_types_vegetation@1" = 1
Save the Output layer as: forets.tif
The result is a binary layer (0 and 1) where value 1 corresponds to forests.
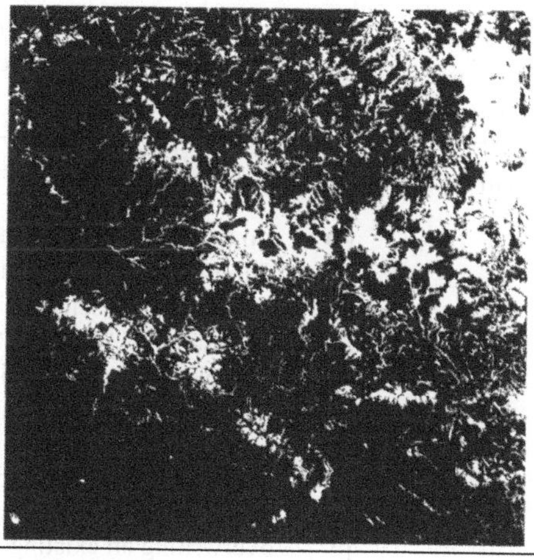</td></tr>
<tr><td>2. Create the binary masks for "Shrublands and herbaceous vegetation" and "Bare soils and sparse vegetation"</td><td>In the Raster calculator:
• Enter the following expression:
"grands_types_vegetation@1" = 2
Save the Output layer as: vegetation_abrustive.tif
In the Raster calculator:
• Enter the following expression:
"grands_types_vegetation@1" = 3
Save the Output layer as: vegetation_eparse.tif
At the end of this step, three raster masks were created corresponding to each large vegetation type.</td></tr>
</table>

Table 6.17. *Creation of large vegetation types binary masks*

6.3.4.3. *Zonal statistics of the large vegetation types' masks in the PPE polygons*

The estimates of the percentage of each large vegetation class in each PPE polygon is carried out by zonal statistics between the binary masks and the PPE vector layer. Indeed, the mean of the values of the binary masks' pixels, whose possible values are 0 (absence) and 1 (presence), is equivalent to the calculation of the percentage of the mask values for each polygon.

Approach	Practical work in QGIS
1. Open the large vegetation types' masks and the PPE vector file	In **QGIS**: • Open the binary masks previously created: forets.tif vegetation_abrustive.tif vegetation_eparse.tif • Open the PPE vector file: DAVAR_PPE_NC_SELECTION_SIMPLIFIE.shp
2. Zonal statistics	In the **menu bar**: • Go to **Raster > Zonal statistics > Zonal statistics** • Choose the raster layer forets.tif • Verify that the selected band is the band 1 • Choose DAVAR_PPE_NC_SELECTION_SIMPLIFIE as the polygon layer containing the zones • Enter "GTV_FO" as Output column prefix. The calculation of zonal statistics creates new fields in the vector attribute table. the choice of the prefix must be clear enough to quickly know the content of the field and be unique to avoid overwriting an existing field. • Select "Mean" as "Statistics to calculate":

<table>
<tr>
<td></td>
<td>
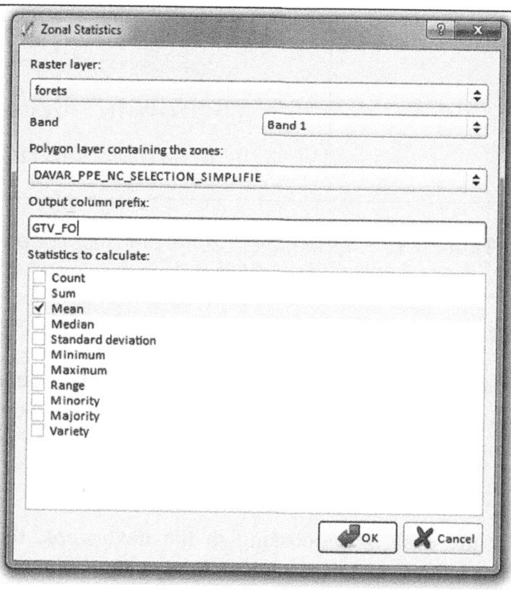

Click OK.
</td>
</tr>
<tr>
<td>3. Zonal statistics check</td>
<td>
Check the results of the Zonal statistics calculation by selecting the layer DAVAR_PPE_NC_SELECTION_SIMPLIFIE.shp, right-click and Open the attribute table.

A new field was created named "GTV_FOmean"
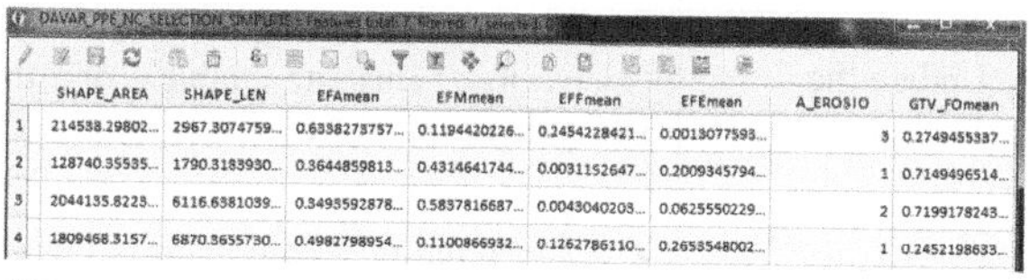
<table>
<tr><th></th><th>SHAPE_AREA</th><th>SHAPE_LEN</th><th>EFAmean</th><th>EFMmean</th><th>EFFmean</th><th>EFEmean</th><th>A_EROSIO</th><th>GTV_FOmean</th></tr>
<tr><td>1</td><td>214538.29802...</td><td>2967.3074759...</td><td>0.6338273757...</td><td>0.1194420226...</td><td>0.2454228421...</td><td>0.0013077593...</td><td>3</td><td>0.2749455337...</td></tr>
<tr><td>2</td><td>128740.35535...</td><td>1790.3183930...</td><td>0.3644859813...</td><td>0.4314641744...</td><td>0.0031152647...</td><td>0.2009345794...</td><td>1</td><td>0.7149496514...</td></tr>
<tr><td>3</td><td>2044135.8223...</td><td>6116.6381039...</td><td>0.3493592878...</td><td>0.5837816687...</td><td>0.0043040203...</td><td>0.0625550229...</td><td>2</td><td>0.7199178243...</td></tr>
<tr><td>4</td><td>1809468.3157...</td><td>6870.3655730...</td><td>0.4982798954...</td><td>0.1100866932...</td><td>0.1262786110...</td><td>0.2653548002...</td><td>1</td><td>0.2452198633...</td></tr>
</table>
</td>
</tr>
<tr>
<td>4. Calculate the zonal statistics for each larger vegetation types' masks</td>
<td>
Repeat the operation with the two other large vegetation types' of masks:

vegetation_abrustive.tif
vegetation_eparse.tif

by naming the Output column prefix, respectively, as follows: GTV_AB and GTV_EP

At the end of this step, the vector layer file has three additional fields: GTV_FOmean, GTV_ABmean and GTV_EPmean
</td>
</tr>
</table>

GTV_FOmean	GTV_ABmean	GTV_EPmean
0.2749455337...	0.1603485838...	0.1311546840...
0.7149496514...	0.2850503485...	0
0.7199178243...	0.2725004891...	0.0007826257...

Table 6.18. *Calculation of zonal statistics for each large vegetation type*

6.3.4.4. *Dominant vegetation in each PPE determination*

This step consists of applying the established decision rule to classify each PPE according to the percentage of each large vegetation type. This processing is performed by calculating a new attribute from three fields calculated in the previous step.

For the ease of processing in the next steps, the PPE dominant vegetation is coded as a discontinuous integer whose values are:

– 1: forests;

– 2: shrublands and herbaceous vegetation;

– 3: bare soils and sparse vegetation.

Approach	Practical work in QGIS
1. Update the attribute table	Open the attribute table of the layer DAVAR_PPE_ NC_SELECTION_SIMPLIFIE.shp • Open the field calculator in clicking on In the **Field calculator**: • Select **Create a new field** • Enter the **Output field name**: T_PAYSAG • Select the **Output field Type** as: Whole number (integer) • Specify the **Output field length** = 10 • Enter the following condition in the **Expression tab**: CASE WHEN (("GTV_FOmean">0.59) OR ("GTV_ABmean">0.33 AND "GTV_FOmean"<=0.59 AND

<table>
<tr>
<td></td>
<td>"GTV_EPmean"<=0.11)

) THEN 1

WHEN (

"GTV_FOmean"<=0.59 AND "GTV_EPmean">0.11 AND "GTV_EPmean"<0.36

) THEN 2

WHEN (

"GTV_FOmean"<0.59 AND "GTV_EPmean">0.36

) THEN 3

ELSE 0

END</td>
</tr>
<tr>
<td></td>
<td>• Click OK

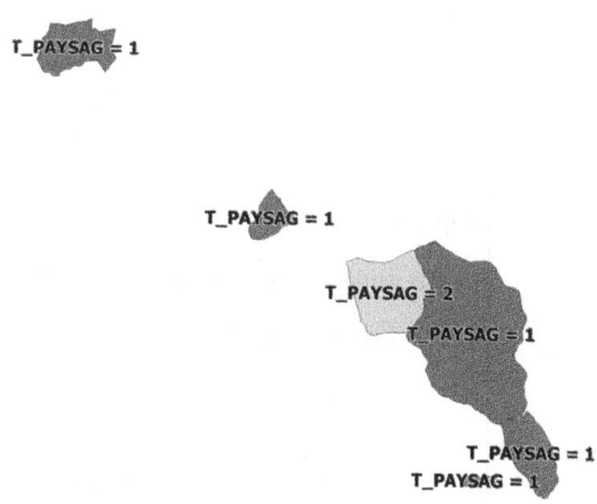

At the end of this step, the field T_PAYSAG corresponds to the dominant vegetation type characterizing each PPE polygons as 3: PPE where bare soils and sparse vegetation dominate, 2: PPE where shrublands and herbaceous vegetation dominate, 1: PPE where forests dominate.</td>
</tr>
</table>

Table 6.19. *Determination of the dominant vegetation in each PPE. For a color version of the table, see www.iste.co.uk/baghdadi/qgis3.zip*

6.3.5. *Step 4: forest fragmentation parameter*

Forest fragmentation determination is realized for each PPE. This involves the selection and intersection of forest polygons with the PPE polygons.

The individual parameters for each forest (perimeter, area) are calculated by polygon. Group statistics make it possible to estimate the two fragmentation indices from these attributes for each PPE.

Finally, the fragmentation parameter is then estimated by applying the decision rule establish in section 6.2.4.4. for each PPE.

6.3.5.1. *Forest vector layer intersection with PPE polygons*

The first step is to intersect the polygons corresponding to forest from the large vegetation type vector file with the PPE polygons.

Approach	Practical work in QGIS
1. Open the vector files	In **QGIS**: • Open the files: o occupation_sol_2008_spot5_objet_decoupe.shp o DAVAR_PPE_NC_SELECTION_SIMPLIFIE.shp
2. Select the "Forest" polygons and create a new vector file.	Open the attribute table of the file: occupation_sol_2008_spot5_objet_decoupe.shp In the **attribute table**: • Select all the polygons where the filed CLASSE = residential areas by clicking on 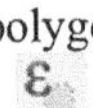 o In the field **Expression**, enter the condition: "GT_VEG" = 1 o Click on **Select** Select the layer: occupation_sol_2008_spot5_objet_decoupe.shp And click-right, **Save as...** • Name the Output file as forets.shp; • Select the option "Save only selected features" • Click OK.

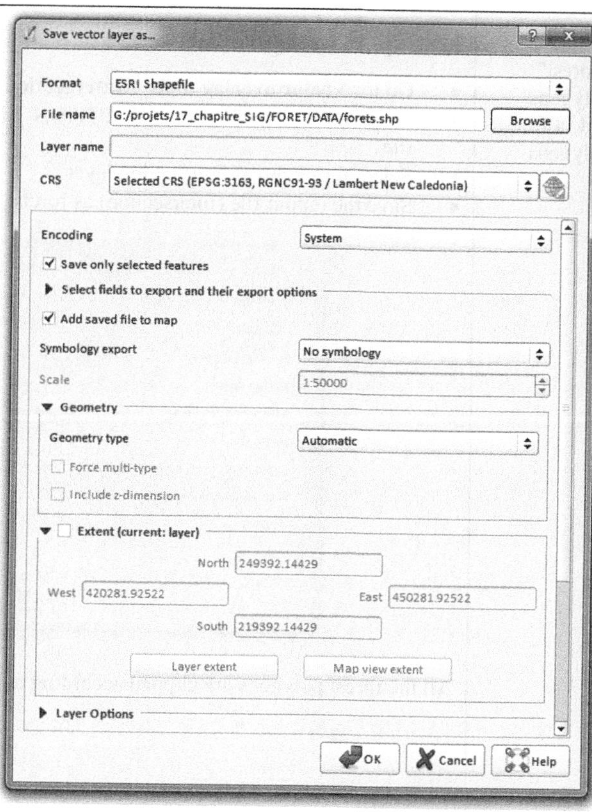

The result is a vector layer containing only "forests" polygons.

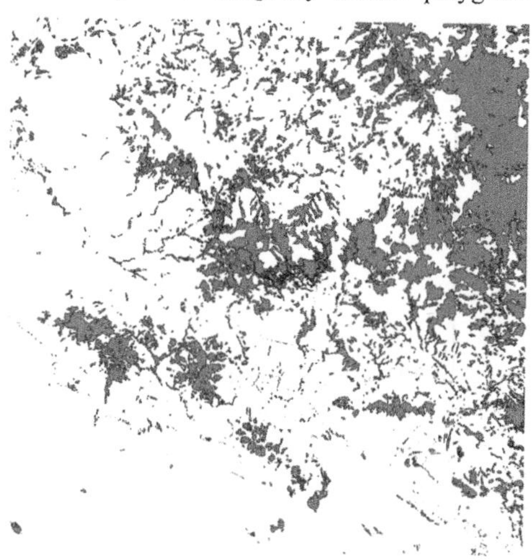

<table>
<tr>
<td>3. Intersect "Forest" polygons and PPE polygons</td>
<td>
In the QGIS Geoalgorithms:

- Go to: **Vector overlay tools > Intersection**
- Select DAVAR_PPE_NC_SELECTION_SIMPLIFIE.shp as input file
- Select forets.shp as intersection layer
- Save the output file (Intersection) as forets_ppe_simplifie.shp

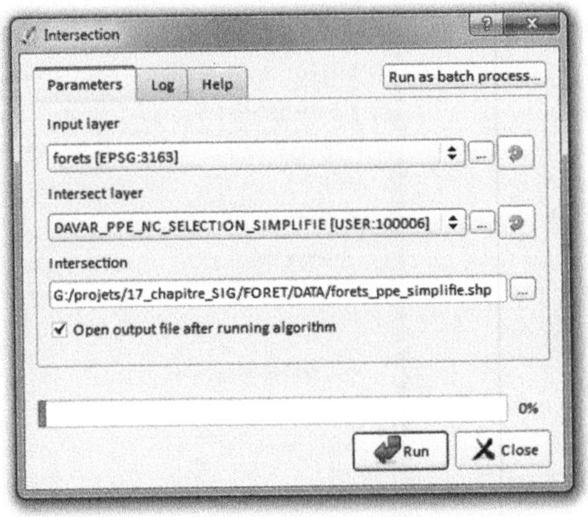

All the forest polygons are clipped according to PPE polygons.

The intersection process also aggregates the attributes of the two vectors. Thus, the membership of each forest polygon intersected with a PPE is provided by the field ID_PPE.
</td>
</tr>
</table>

Table 6.20. *Clipping forests in the PPE. For a color version of the table, see www.iste.co.uk/baghdadi/qgis3.zip*

6.3.5.2. *Forest fragmentation indices calculation*

The calculation of the selected fragmentation indices (shape and diversity, equations [6.1] and [6.2]) is based on the geometric parameters "area" and "perimeter" of the polygons.

Approach	Practical work in QGIS
1. Computer perimeters	Open the attribute table of the file: forets_ppe_simplifie.shp • open the field calculator in clicking in In the **Field calculator**: • Select **Create a new field** • Enter the **Output field name**: perim • Select the **Output field type**: Decimal number(real) • Enter **Output field length** = 12 • And enter **Precision** = 6 (number of digits after the decimal point) • In the **Expression tab**, enter the following condition: $perimeter • Click OK
2. Computer areas	In the **Field calculator**: • Select **Create a new field** • Enter the **Output field name**: area • Select the **Output field type**: Decimal number(real) • Enter **Output field length** = 12 • And enter **Precision** = 6 (number of digits after the decimal point) • In the **Expression tab**, enter the following condition: $area • Click OK
3. Simplify the attribute table	To simplify the attribute table of the vector file, remove the unnecessary fields in this layer. Open the attribute table of the vector file: forets_ppe_simplifie.shp • Click on to delete fields

<table>
<tr><td></td><td>

- We only keep the fields "ID_PPE", "perim" and "area" of this vector file. To delete the other fields, select them in the list of fields by highlighting them:

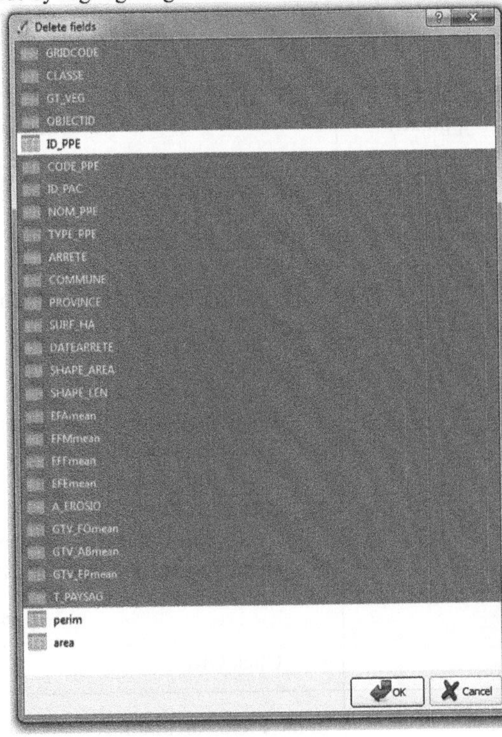

- Click OK

</td></tr>
</table>

Table 6.21. *Calculation of perimeters and areas of forest polygons*

From this file, the sums of areas and squared perimeters by PPE is carried out with the group statistic plugin "Group Stats".

Approach	Practical work in QGIS
1. Open the plugin Group Stats	In the **menu bar**: • Click **Vector** > **Group Stats** > **Group Stats**
2. Calculate group statistics	In the **Control panel** on the right part of the window: • In the field **Layer**, select the layer forets_ppe_simplifie.shp • Select **ID_PPE** and drag and drop it in the section **Rows** at the bottom left of the **control panel**

	• Select **perim** and drag and drop it in the section **Value** at the bottom right of the **control panel** • Select **Sum** and drag and drop it in the section **Value** in the bottom left section of the **control panel** with **perim2** Click on **Calculate** • In the left part of the window, a table displays the sum of the perimeter for each PPE • In the menu bar of Group Stats, select **Data > Save All to CSV File...** • Save the file as somme_perim.csv Calculate the sum of the area grouped by ID_PPE using the same procedure and save the resulting table as somme_area.csv Close Group Stats plugin

Table 6.22. *Calculation of group statistics by PPE*

To calculate the diversity index, we must calculate for each polygon i of the class "forest" j the component D_{ij} as:

$$D_{ij} = -\left(\frac{a_{ij}}{\sum_{i=1}^{n_j} a_{ij}} ln \frac{a_{ij}}{\sum_{i=1}^{n_j} a_{ij}}\right). \qquad [6.3]$$

From equation [6.1], the diversity index for each PPE becomes:

$$H_j(a) = \sum_{i=1}^{n_j} D_{ij}. \qquad [6.4]$$

The table of the sum of areas created previously is joined to the file forets_ppe_simplifie.shp using their common field ID_PPE in order to continue the calculation of the fragmentation indices.

Approach	Practical work in QGIS
1. Open a spreadsheet layer in QGIS	• In **QGIS**, open a spreadsheet table in clicking on : o Select *.* file type as input file o Open the file somme_area.csv o Automatically, the import window displays the import settings o Modify the name of the layer as "somme_aire" o Change the attribute type ID_PPE to Integer o Change the attribute none to real o Click OK

<table>
<tr>
<td></td>
<td>
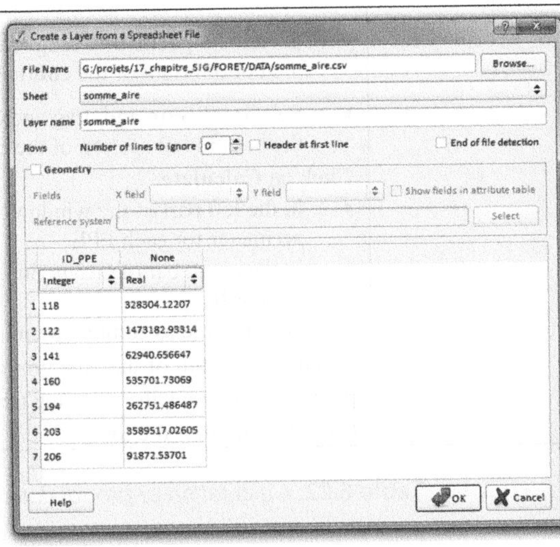

<ul><li>The table appears in the layers panel of QGIS</li></ul>

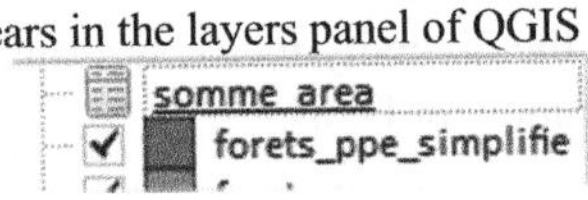

</td>
</tr>
<tr>
<td>2. Join a spreadsheet table to a vector layer</td>
<td>
Select the layer forets_ppe_simplifie.shp and right-click > properties

<ul>
<li>Go to the Joins tab </li>
<li>Click  to add a table join.</li>
<li>In the window Add vector join:
<ul>
<li>Select the spreadsheet layer somme_area as join layer</li>
<li>Select the attribute ID_PPE as Join field</li>
<li>Select ID_PPE as Target field</li>
<li>Select the option Cache join layer in virtual memory</li>
<li>Select the option Choose which fields are joined and select the field "none"</li>
<li>Select Custom field name prefix and leave somme_area_</li>
</ul>
</li>
<li>Click OK</li>
</ul>

At the end of this step, the layer forets_ppe_simplifie.shp has a new field called "somme_area_None"

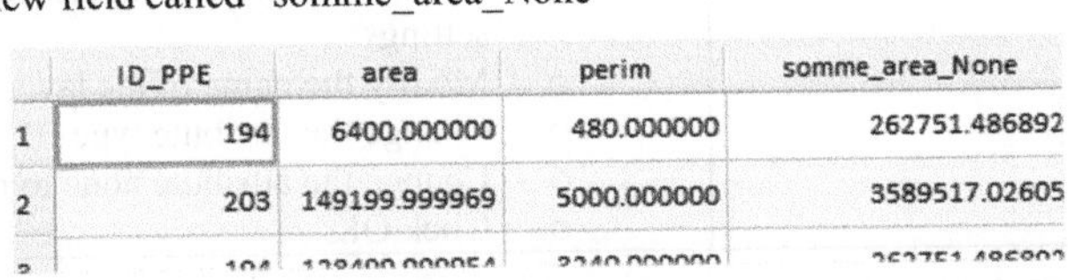

</td>
</tr>
</table>

3. Calculate the D component (equation [6.3]) for each "forest" polygon	Open the attribute table of the layer forets_ppe_simplifie.shp • Open the field calculator in clicking in [icon] In the **Field calculator**: • Select **Create a new field** • Enter the **Output field name**: D • Select the **Output field type**: Decimal number(real) • Enter **Output field length** = 12 • And enter **Precision** = 6 (number of digit after the decimal point) • In the **Expression tab**, enter the following condition: -("area" / "somme_area_None")* ln("area" / "somme_area_None") • Click OK

Table 6.23. *D component calculation*

Diversity index is the sum of the D component for each PPE according to equation [6.4]. This calculation is performed using the Group Stats plugin.

Approach	Practical work in QGIS
1. Open Group Stats plugin	In the **menu bar**: • Click **Vector > Group Stats > Group Stats**
2. Calculate the Group Statistics	In the **Control panel** on the right of the window: • In the field **Layer**, select the layer forets_ppe_simplifie.shp • Select [icon] **ID_PPE** and drag and drop it in the section **Rows** at the bottom left of the **control panel** • Select [icon] **D** and drag and drop it in the section **Value** at the bottom right of the **control panel** • Select [icon] **Sum** and drag and drop it in the section **Value** in the bottom left section of the **control panel** with ***D*** Click on **Calculate** • In the left part of the window, a table displays the sum of the D component for each PPE • In the menu bar of Group Stats, select **Data > Save All to CSV File...**

<table>
<tr><td></td><td>

- Save the file as nomindice_diversite.csv

The result is a CSV table containing the diversity index for each PPE (ID_PPE). The diversity index ranges from 0.04 to 2.35.

<table>
<tr><th>ID_PPE</th><th></th></tr>
<tr><td>118</td><td>0,667209</td></tr>
<tr><td>122</td><td>0,070431</td></tr>
<tr><td>141</td><td>0,168543</td></tr>
<tr><td>160</td><td>0,809199</td></tr>
<tr><td>194</td><td>1,45977</td></tr>
<tr><td>203</td><td>2,3555</td></tr>
<tr><td>206</td><td>0,040689</td></tr>
</table>

Close Group Stats plugin.
</td></tr>
</table>

Table 6.24. *Diversity index calculation*

The diversity index is joined to the PPE vector layer.

<table>
<tr><th>Approach</th><th>Practical work in QGIS</th></tr>
<tr><td>1. Open a spreadsheet layer in QGIS</td><td>

- In **QGIS**, open a spreadsheet table in clicking on :
  - Select *.* file type as input file
  - Open the file indice_diversite.csv
  - Automatically, the import window displays the import settings
  - Modify the name of the layer as "diversite"
  - Change the attribute type ID_PPE to Integer
  - Change the attribute none to real
  - Click OK
</td></tr>
</table>

<table>
<tr>
<td></td>
<td>
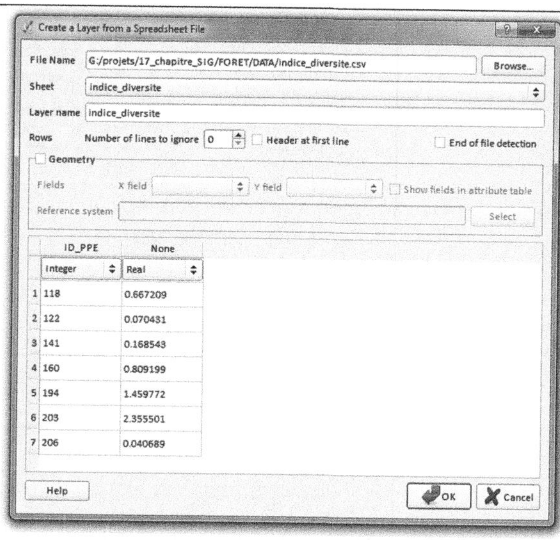

<ul><li>The table appears in the layers panel of QGIS:</li></ul>

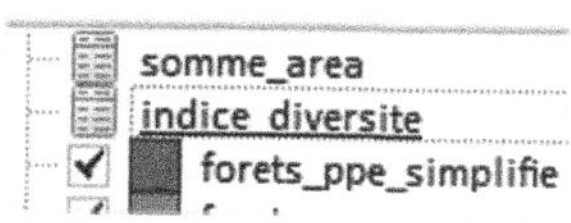

</td>
</tr>
<tr>
<td>2. Join a spreadsheet table to a vector layer</td>
<td>
Select the layer DAVAR_PPE_NC_SELECTION_SIMPLIFIE.shp and right-click > properties
<ul>
<li>Go to the Joins tab </li>
<li>Click  to add a table join.</li>
<li>In the window Add vector join:
<ul>
<li>Select the spreadsheet layer indice_diversite as join layer</li>
<li>Select the attribute ID_PPE as Join field</li>
<li>Select ID_PPE as Target field</li>
<li>Select the option Cache join layer in virtual memory</li>
<li>Select the option Choose which fields are joined and select the field "none"</li>
<li>Select Custom field name prefix and leave indice_diversite_</li>
</ul>
</li>
<li>Click OK</li>
</ul>
</td>
</tr>
</table>

	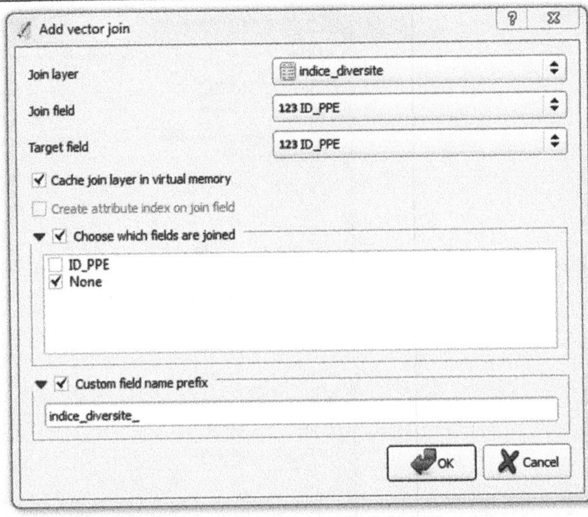 At the end of this step, the layer DAVAR_PPE_NC_SELECTION_ SIMPLIFIE.shp has a new field called "indice_diversite_None". 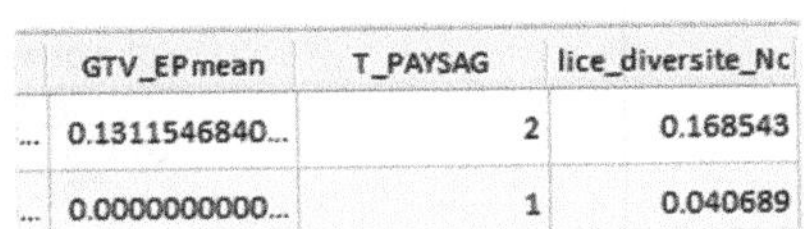

Table 6.25. *Join the table of the diversity index to the vector layer of PPE*

According to equation [6.2], the shape index for each PPE is the ratio between the sum of the square perimeters of forest patches and the sum of the square forest patches' areas. It is therefore computed directly in the PPE vector layer from the join field of the sum of the square perimeters and the field areas.

Approach	Practical work in QGIS
1. Open a spreadsheet layer in QGIS	• In **QGIS**, open a spreadsheet table in clicking on  : o Select *.* file type as input file o Open the file somme_perim.csv o Automatically, the import window displays the import settings o Modify the name of the layer by "somme_perim" o Change the attribute type ID_PPE to Integer o Change the attribute none to real o Click OK

<table>
<tr>
<td></td>
<td>
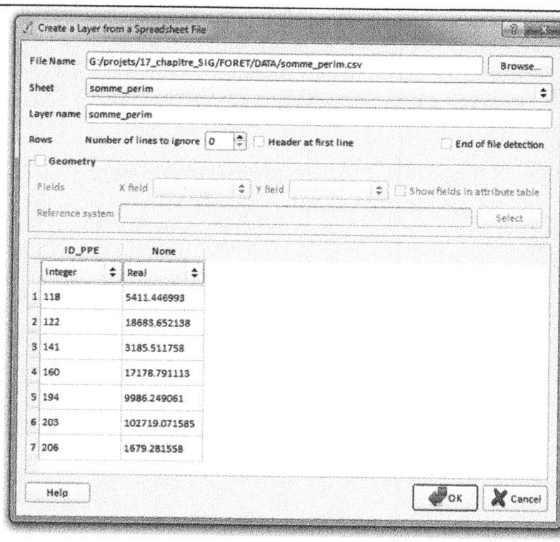

<ul><li>The table appears in the layers panel of QGIS</li></ul>
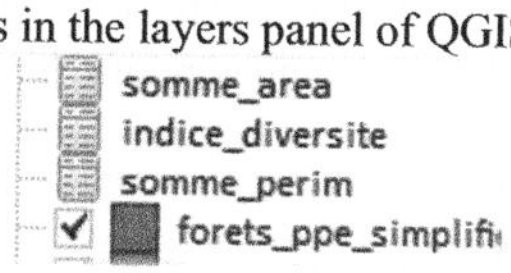

</td>
</tr>
<tr>
<td>2. Join a spreadsheet table to a vector layer</td>
<td>
Select the layer DAVAR_PPE_NC_SELECTION_SIMPLIFIE.shp and right-click > properties
<ul>
<li>Go to the Joins tab </li>
<li>Click  to add a table join</li>
<li>In the window Add vector join:
<ul>
<li>Select the spreadsheet layer somme_perim as join layer</li>
<li>Select the attribute ID_PPE as Join field</li>
<li>Select ID_PPE as Target field</li>
<li>Select the option Cache join layer in virtual memory</li>
<li>Select the option Choose which fields are joined and select the field "none"</li>
<li>Select Custom field name prefix and leave somme_perim_</li>
</ul>
</li>
<li>Click OK</li>
</ul>
</td>
</tr>
</table>

<table>
<tr>
<td></td>
<td>
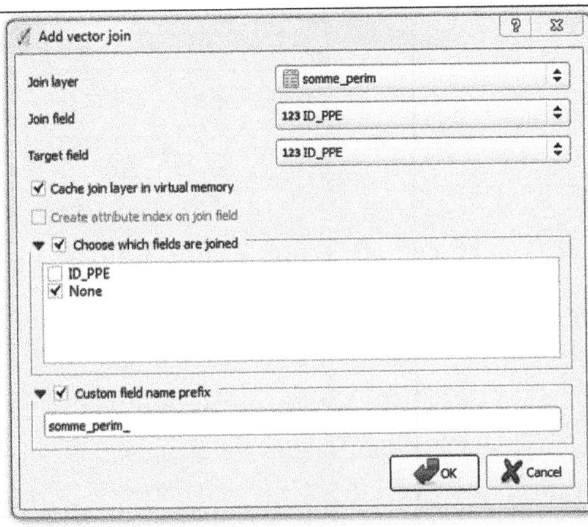

At the end of this step, the layer DAVAR_PPE_NC_SELECTION_SIMPLIFIE.shp has a new field called "somme_perim_None"

<table>
<tr><th></th><th>GTV_EPmean</th><th>T_PAYSAG</th><th>indice_diversite_None</th><th>somme_perim_None</th></tr>
<tr><td>...</td><td>0.1311546840...</td><td>2</td><td>0.168543</td><td>3185.511758</td></tr>
<tr><td>...</td><td>0.0000000000...</td><td>1</td><td>0.040689</td><td>1679.281558</td></tr>
</table>

Repeat the operation with the table somme_area.
</td>
</tr>
<tr>
<td>3. Calculate the shape index</td>
<td>
Open the attribute table of the file:
DAVAR_PPE_NC_SELECTION_SIMPLIFIE.shp
• Open the field calculator in clicking in 
In the Field calculator:
• Select Create a new field
• Enter the Output field name: forme
• Select the Output field type: Decimal number(real)
• Enter Output field length = 12
• And enter Precision = 6
(number of digits after the decimal point)
• In the Expression tab, enter the following condition:
("somme_perim_None" * "somme_perim_None") / "somme_area_None"
• Click OK
</td>
</tr>
</table>

Table 6.26. *Shape index calculation*

At the end of this step, the diversity index (field "indice_diversite_None") and the shape index (field "forme") are calculated for each PPE.

6.3.5.3. *Forest fragmentation estimate for each PPE*

This step consists of applying the established decision rule to classify each PPE according to the threshold values defined for each fragmentation index. This processing is performed by calculating a new attribute from two fields calculated in the previous step.

For the ease of processing in the next steps, the PPE forest fragmentation is coded as a discontinuous integer whose values are:

– 1: high to very high forest fragmentation;

– 2: moderate to high forest fragmentation;

– 3: low to moderate forest fragmentation.

Approach	Practical work in QGIS
1. Calculate the forest fragmentation parameter	Open the attribute table of the file: DAVAR_PPE_NC_SELECTION_SIMPLIFIE.shp • Open the field calculator in clicking in In the **Field calculator**: • Select **Create a new field** • Enter the **Output field name**: I_FRAG • Select the **Output field type**: Whole number (integer) • Enter **Output field length** = 10 • In the **Expression tab**, enter the following condition: CASE WHEN ("indice_diversite_None">1.9 AND "forme">=400) THEN 1 WHEN (("indice_diversite_None">1.9 AND "forme"<400) OR ("indice_ diversite_None">=0.6 AND "indice_diversite_None"<=1.9 AND "forme"> =400) OR ("indice_diversite_None"<0.6 AND "forme">2250)

<table>
<tr><td></td><td>) THEN 2

WHEN (("indice_diversite_None">=0.6 AND "indice_diversite_None"<=1.9 AND "forme"<400) OR ("indice_diversite_None"<0.6 AND "forme"<=2250)

) THEN 3

ELSE 0

END</td></tr>
<tr><td></td><td>• Click OK

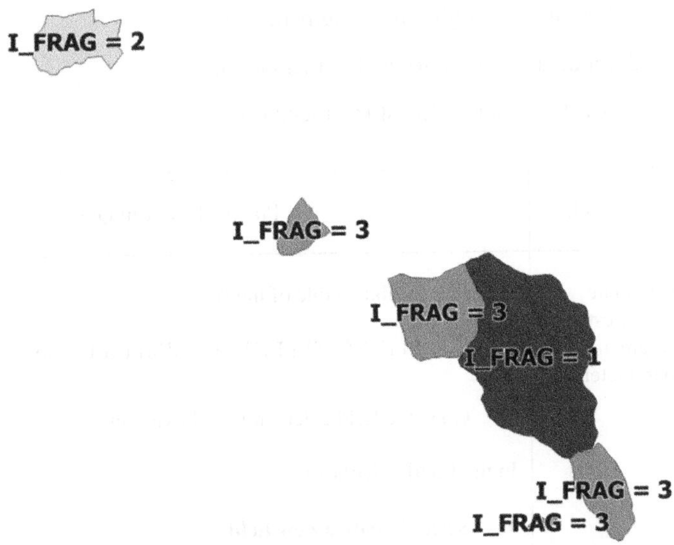

At the end of this step, the field I_FRAG corresponds to the forest fragmentation parameter and distinguishes each PPE in three classes: three for PPE where forests are slightly too moderately fragmented, 2 for PPE where forests are moderately to highly fragmented, and 1 for PPE where forests are highly to very highly fragmented.</td></tr>
</table>

Table 6.27. *Forest fragmentation estimate for each PPE. For a color version of the table, see www.iste.co.uk/baghdadi/qgis3.zip*

6.3.6. *Step 5: forest function assessment for water protection*

Estimating forest function state for water resource protection consists of concatenating the three parameters: erosion hazard, dominant vegetation and forest fragmentation. All the possible combinations are tested according to the expert decision tree defined previously to assign the forest function state.

This assessment is performed in two steps:

– parameters concatenation;

– application of the decision tree.

Approach	Practical work in QGIS
1. Concatenate the 3 parameters: erosion hazard, dominant vegetation, forest fragmentation	Concatenation consists of creating an attribute that aggregates the value of each parameter together in the form ABC, with A: erosive hazard, B: landscape tendency and C: forest fragmentation. Open the attribute table of the file: DAVAR_PPE_NC_SELECTION_SIMPLIFIE.shp • Open the field calculator in clicking in In the **Field calculator**: • Select **Create a new field** • Enter the **Output field name**: CONCAT • Select the **Output field type**: Whole number (integer) • Enter **Output field length** = 10 • In the **Expression tab**, enter the following condition: "A_EROSIO" || "T_PAYSAG" || "I_FRAG" • Click OK

<table>
<tr>
<td></td>
<td>
<table>
<tr><th>CONCAT</th></tr>
<tr><td>323</td></tr>
<tr><td>113</td></tr>
<tr><td>213</td></tr>
<tr><td>112</td></tr>
<tr><td>113</td></tr>
<tr><td>123</td></tr>
<tr><td>111</td></tr>
</table>
At the end of this step, the field CONCAT is in the form of a three-digit number where each digit corresponds to one of the parameter.
</td>
</tr>
<tr>
<td>2. Assess the forest function assessment for each PPE</td>
<td>
This is performed by translating the expert decision tree into an expression exploiting the previously calculated concatenation field.

Open the attribute table of the file:

DAVAR_PPE_NC_SELECTION_SIMPLIFIE.shp

• open the field calculator in clicking in

In the Field calculator:

• Select Create a new field
• Enter the Output field name: FONCTION
• Select the Output field type: Whole number (integer)
• Enter Output field length = 10
• In the Expression tab, enter the following condition:
<hr>
CASE WHEN (

"CONCAT"=331 OR

"CONCAT"=332 OR

"CONCAT"=333 OR

"CONCAT"=221 OR

"CONCAT"=232 OR

"CONCAT"=231 OR

"CONCAT"=112 OR

"CONCAT"=111 OR
</td>
</tr>
</table>

<table>
<tr><td></td><td>"CONCAT"=122 OR
"CONCAT"=121 OR
"CONCAT"=132 OR
"CONCAT"=223 OR
"CONCAT"=222 OR
"CONCAT"=233 OR
"CONCAT"=113 OR
"CONCAT"=123 OR
"CONCAT"=133 OR
"CONCAT"=131
) THEN 1
WHEN (
"CONCAT"=312 OR
"CONCAT"=311 OR
"CONCAT"=322 OR
"CONCAT"=211 OR
"CONCAT"=321 OR
"CONCAT"=212 OR
"CONCAT"=213 OR
"CONCAT"=323
)THEN 2
WHEN (
"CONCAT"=313
)THEN 3
ELSE 0
END
• Click OK</td></tr>
</table>

<table>
<tr>
<td></td>
<td>
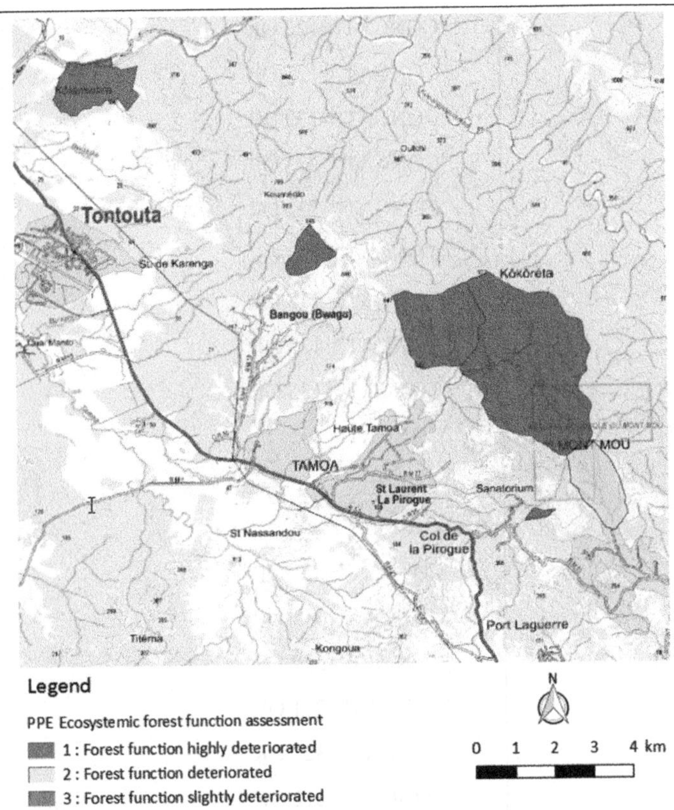

At the end of this step, the field FUNCTION corresponds to the forest function to protect the water resource state assessment for each PPE with 3: PPE where the function of the forests is highly deteriorated; 2: PPE where the function of the forests is deteriorated; 1: PPE where the function of the forests is slightly deteriorated.
</td>
</tr>
</table>

Table 6.28. *Forest function state assessment. For a color version of the table, see www.iste.co.uk/baghdadi/qgis3.zip*

6.4. Bibliography

[AND 16] ANDREOLI R., CIESLAK J.D., VAN HAAREN B. *et al.*, Diagnostic de la couverture forestière et des services écosystémiques des Périmètres de Protection Eloignée des Captages d'Eau sur la Grande Terre et l'Ile des Pins en Nouvelle-Calédonie, BLUECHAM SAS/WWF, Report, 2016.

[AND 18] ANDREOLI R., "Modeling erosion risk using the RUSLE equation", in BAGHDADI N., MALLET C., ZRIBI M. (eds), *QGIS and Applications in Water and Risks*, ISTE, London and John Wiley & Sons, New York, 2018.

[BOG 05] BOGAERT J., MAHAMANE A., "Ecologie Du Paysage: Cibler la Configuration et l'échelle Spatiale," *Annales Des Sciences Agronomiques*, vol. 7, no. 1, pp. 39–68, 2005.

[BOG 00] BOGAERT J., ROUSSEAU R., VAN HECKE P. *et al.*, "Alternative Area-Perimeter Ratios for Measurement of 2D Shape Compactness of Habitats", *Applied Mathematics and Computation*, vol. 111, no. 1, pp. 71–85, 2000.

[BRA 07a] BRADSHAW C.J., SODHI N.S., PEH K.S. *et al.*, "Global evidence that deforestation amplifies flood risk and severity in the developing world", *Global Change Biology*, vol. 13, no. 11, pp. 2379–2395, 2007.

[BRA 07b] BRAUMAN K.A., DAILY G.C., DUARTE K. *et al.*, "The nature and value of ecosystem services: an overview highlighting hydrologic services", *Annual Review of Environment and Resources,* vol. 32, pp. 67–98, 2007.

[BRO 16] BROOKHUIS B.J., HEIN L.G., "The value of the flood control service of tropical forests: a case study for Trinidad", *Forest Policy and Economics*, vol. 62, pp. 118–124, 2016.

[BRO 02] BROOKS T.M., MITTERMEIER R.A., MITTERMEIER C.G. *et al.*, "Habitat loss and extinction in the hotspots of biodiversity", *Conservation Biology*, vol. 16, no. 4, pp. 909–923, 2002.

[CUR 15] CURT T., BORGNIET L., IBANEZ T., *et al.*, "Understanding fire patterns and fire drivers for setting a sustainable management policy of the New-Caledonian biodiversity hotspot", *Forest Ecology and Management*, vol. 337, pp. 48–60, 2015.

[DTS 08] DTSI, "Notice de La Classification de L'occupation Du Sol de La Nouvelle Calédonie Par Approche Objet V1.0 – 2008", available at http://sig-public.gouv.nc/Notice-Occupationdusol2008-SPOT5-approcheobjet.pdf, 2008.

[FAO 99] FAO, "Manuel de Statistique Pour La Recherche Forestière", available at http://www.fao.org/docrep/003/X6831F/X6831f00.htm, 1999.

[FOL 07] FOLEY J.A., ASNER G.P., COSTA M.H. *et al.*, "Amazonia revealed: forest degradation and loss of ecosystem goods and services in the Amazon Basin", *Frontiers in Ecology and the Environment,* vol. 5, no. 1, pp. 25–32, 2007.

[GUI 05] GUILLEMETTE F., PLAMONDON A.P., PREVOST M. *et al.*, "Rainfall generated stormflow response to clearcutting a boreal forest: peak flow comparison with 50 world-wide basin studies", *Journal of Hydrology,* vol. 302, nos. 1–4, pp. 137–153, 2005.

[GYS 05] GYSSELS G., POESEN J., BOCHET E. *et al.*, "Impact of plant roots on the resistance of soils to erosion by water: a review", *Progress in Physical Geography,* vol. 29, no. 2, pp. 189–217, 2005.

[HUR 02] HURD J.D., WILSON E.H., CIVCO D.L., "Development of a forest fragmentation index to quantify the rate of forest change", ASPRS-ACSM Annual Conference and FIG XXII Congress, Washington, DC, 2002.

[IBA 13] IBANEZ T., BORGNIET L., MANGEAS M., *et al.*, "Rainforest and Savanna landscape dynamics in New Caledonia: towards a mosaic of stable rainforest and Savanna states?", *Austral Ecology*, vol. 38, no. 1, pp. 33–45, 2013.

[JAF 80] JAFFRÉ T., "Étude écologique du peuplement végétal des sols dérivés de roches ultrabasiques en Nouvelle Calédonie", *Trav. et doc. de l'ORSTOM*, no. 124, p. 274, 1980.

[KEI 03] KEIM R.F., SKAUGSET A.E., "Modeling effects of forest canopies on slope stability", *Hydrological Processes*, vol. 17, no. 7, pp. 1457–1467, 2003.

[KLE 89] KLEIN B.C., "Effects of Forest Fragmentation on Dung and Carrion Beetle Communities in Central Amazonia", *Ecology*, vol. 70, no. 6, pp. 1715–1725, 1989.

[LAC 17] LACAZE B., DUDEK J., PICARD J, *GRASS GIS Software with QGIS*, vol. 1, 2017.

[LAU 98] LAURANCE W.F., FERREIRA L.V., RANKIN-DE MERONA J.M. *et al.*, "Rain forest fragmentation and the dynamics of Amazonian tree communities", *Ecology*, vol. 79, no. 6, pp. 2032–2040, 1998.

[LE 02] LE BISSONAIS Y., THORETTE J., BARDET C. *et al.*, L'érosion hydrique des sols en France, IFEN/INRA, Report, p. 106, 2002.

[LE 04] LE BISSONNAIS Y., DUBREUILN., DAROUSSIN J. *et al.*, "Modélisation et cartographie de l'aléa d'érosion des sols à l'échelle régionale", *Etude et gestion des sols*, vol. 11, no. 3, pp. 307–321, 2004.

[MAR 11] MARESCA B., MORDRET X., UGHETTO A.L. *et al.*, "Évaluation des services rendus par les écosystèmes en France: Les enseignements d'une application du Millennium Ecosystem Assessment au territoire français", *Développement durable et territoires*, http://developpementdurable.revues.org/9053, vol. 2, no. 3, 2011.

[MC 99] MC COY S., JAFFRÉ T., RIGAULT F. *et al.*, "Fire and Succession in the Ultramafic Maquis of New Caledonia", *Journal of Biogeography*, vol. 26, no. 3, pp. 579–594, 1999.

[MCG 95] MCGARIGAL K., MARKS B.J., FRAGSTATS: spatial pattern analysis program for quantifying landscape structure, Technical Report, PNW-GTR-351, p. 122, 1995.

[MIL 03] MILLENNIUM ECOSYSTEM ASSESSMENT, Ecosystems and Human Well-being: Our Human Planet, Report, Washington, DC, 2003.

[MOU 12] MOUHAMADOU I.T., TOURE F., IMOROU I.T., *et al.*, "Indices de structures spatiales des îlots de forêts denses dans la région des Monts Kouffé", *VertigO-la revue électronique en sciences de l'environnement*, http://vertigo.revues.org/13059, vol. 12, no. 3, 2012.

[RAH 01] RAHERILALAO M.J., "Effets de la fragmentation de la forêt sur les oiseaux autour du Parc National de Ranomafana (Madagascar)", *Revue d'écologie*, vol. 56, no. 4, pp. 389–406, 2001.

[RIT 00] RITTERS K., WICKHAM J., O'NEILL R., *et al.*, "Global scale patterns of forest fragmentation", *Conservation Ecology*", available at: http://www.ecologyandsociety.org/vol4/iss2/ art3/, vol. 4, no. 2, 2000

[SMA 01] SMAKHTIN V.U., "Low flow hydrology: a review", *Journal of Hydrology*, vol. 240, no. 3, pp. 147–186, 2001.

7

Assessing the Influence of Landscape on Biodiversity Using the QGIS Plugin LecoS

7.1. Introduction

We propose in this chapter to illustrate the implementation of a conventional methodology in landscape ecology. The method is based on an existing landscape representation, i.e. a land cover description of the landscape which consists of the following steps: (1) computation of a set of features and (2) use of the features as explanatory variables in a descriptive model built on the observed biodiversity. First, we recall the principles of this approach. Then, we illustrate its partial implementation in QGIS with the LecoS plugin. The final stage of statistical landscape-type modeling is carried out with the R software.

7.2. Principle of the approach

Landscape ecology is a discipline that focuses on understanding the interactions between the spatial and temporal heterogeneity of landscapes and environmental processes [FOR 86, TUR 01]. In this discipline, the landscape is considered to be a space, with variable dimensions, consisting of patches (the habitat), corridors (connections between the patches that allow the species to move from one habitat to another) and a matrix (the frame background, less favorable to the species). This patch–corridor–matrix model, inspired by island biogeography theory [MCA 67], subsequently evolved by relaxing the dichotomy between favorable and unfavorable spaces to the species [SIR 16]. In general, the landscape can be modeled as a mosaic

Chapter written by Sylvie LADET, David SHEEREN, Pierre-Alexis HERRAULT and Mathieu FAUVEL.

of patches of various types, including spatial heterogeneity, which have varying influence on the ecological processes [DUE 97, SIR 17].

Spatial heterogeneity refers to the characterization of two landscape states: the composition and the configuration [FAH 11]. The composition refers to the nature of the patches and their relative abundance in the landscape (e.g. proportion of forests). Configuration expresses the spatial organization of the patches that can be characterized through various indicators such as the level of aggregation or isolation of patches, their spatial distribution, or even the complexity of their shape. Two landscapes may have an identical composition but a different configuration (and vice versa), which is likely to modulate the effects of the landscape on biodiversity (Figure 7.1).

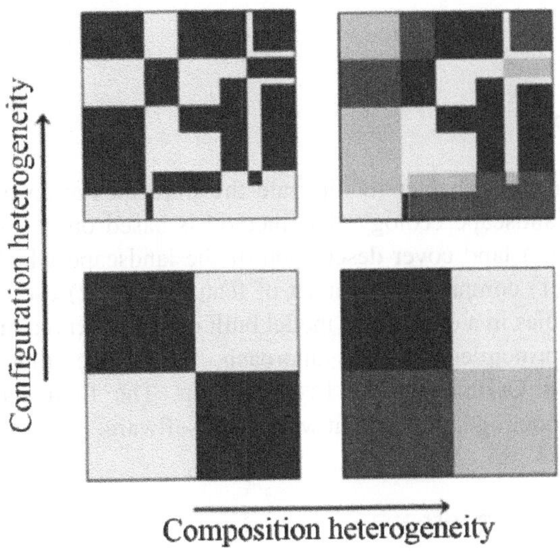

Figure 7.1. *The two components of spatial heterogeneity: composition and configuration. The heterogeneity of composition increases with the number of patches of various types (from left to right). The heterogeneity of configuration increases with the complexity of the organization of the patches (from bottom to top) (source: adapted from [FAH 11]). For a color version of the figure, see www.iste.co.uk/baghdadi/qgis3.zip*

To measure the influence of landscape heterogeneity on biodiversity, a 3-step geomatic approach is thus adopted: (1) production of a landscape representation, (2) computation of landscape descriptors characterizing heterogeneity and (3) statistical modeling of the relationship between indicators of biodiversity and landscape descriptors. It is the translation of a landscape species analysis seen by ecologist researchers (Figure 7.2).

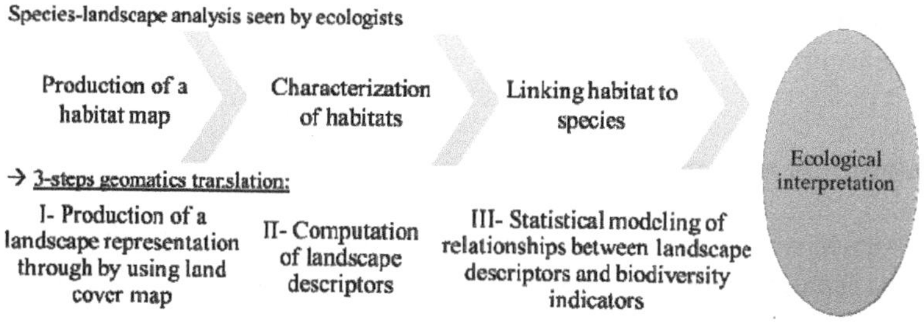

Figure 7.2. *Landscape ecology analysis approach and translation for geomatics processing chain in three steps*

The first step is to build a landscape representation according to the model of a mosaic of patches. In practice, ecologists often rely on land cover maps whose nomenclature is supposed to be adapted to the species of interest (some occupations being assimilated to habitat). This map can be produced by photointerpretation or classification from a set of satellite images [ING 16]. Particular attention should be paid at this stage because the choice of map, with its level of spatial detail and its nomenclature (definition of classes, number of terms) can influence the landscape analysis and measured outcomes.

The second step is calculation of the landscape descriptors (or metrics) characterizing the spatial organization of the patches of habitats. These features can be estimated at three levels of analysis: (1) the patches themselves (e.g. surface, compactness); (2) their class, all patches of the same type (e.g. average perimeter); (3) the landscape, the entire mosaic of the patches in the extent of study site (e.g. diversity index). For each of the levels, there are a large number of possible metrics whose selection and interest may vary according to the studied taxon [GUS 98, UUE 09]. The choice of metrics is thus left to the analyst, according to knowledge on the species occurrence but also according to the correlations that may exist between certain metrics [RII 95]. Limiting their number is sought to build landscape species models [CUS 08].

The last step is the statistical modeling of the relationship between the calculated metrics and indicators of biodiversity. The latter generally are the taxonomic species diversity (i.e. species richness) or abundance. Species can also be grouped into communities, for example, specialists of certain habitats such as the bird species in forest environments. Different techniques of modeling, parametric or not, can then be considered to estimate landscape effects. Conventionally, the generalized linear models (GLMs) or generalized additive models are mobilized [ZUU 09]. However, other more recent machine-learning type approaches can be implemented

[WEG 16]: for example, this is the case for random forests (RFs) also being used in remote sensing.

7.3. Materials and methods

The objective in this chapter is to show the feasibility of this analysis in landscape ecology by combining several free tools interfaced with GIS software:

– the remote sensing tool, *Orfeo Toolbox*[1];

– the landscape descriptors calculation tool, *Landscape ecology statistics* (LecoS)[2];

– the statistical modeling tool, R, and its development interface, R Studio[3].

The proposed three-step method is adapted to the free and open-access software, Quantum GIS (QGIS 2.18 Windows), so that a user, without large expertise in remote sensing and geomatics, can use it without signing up for use of software licenses.

7.3.1. *Step 1: land cover map*

7.3.1.1. *Using an existing land cover map*

It is possible to use land cover or land use maps available in the context of Open Data[4]. There are a few maps of France with various thematic, spatial and temporal resolutions (Table 7.1). The selected product may influence the analysis in terms of spatial scale and detail as well as the land cover classes. These settings may have an impact on the approach to landscape characterization and the measured effects [MOI 16]. Therefore, users should select land cover products with respect to their application.

---

1 Orfeo Toolbox (OTB) URL: https://www.orfeo-toolbox.org/.

2 Landscape ecology statistics (LecoS) URL: https://conservationecology.wordpress.com/qgis-plugins-and-scripts/lecos-land-cover-statistics/.

3 R Development Core Team (2008). R: a language and environment for statistical computing. R Foundation for Statistical Computing, Vienna, Austria. http://www.R-project.org. It is best to start by installing R before installing R Studio: http://www.rstudio.com/products/rstudio/download/.

4 Open Data: Open Data give freedom to use, modify and redistribute the data freely, particularly those financed from public funds. Its goal is to be free and universal, diffusing original public or private data over the Internet (www.ist.agropolis.fr *Collection sheets practices*).

	Corine Land Cover[5]	OCS GE[6]	OSO[7]
Available years	1990/2000/2006/2012	2013	2016
Total number of classes	44	14	17
Update	6 years	3 years	1 year
Minimum collection unit	250,000 m² (25 ha)	500 m² in urban areas and 2,500 m² in rural areas	Pixel of 10 m
Producer	EEA(UE)-Copernicus/ CGDD-SOES	IGN	Theia-Cesbio

Table 7.1. *Comparison of the characteristics of some land cover maps of France*

7.3.1.2. *Building a land cover map*

This part is widely described in other chapters of this book; we have therefore decided to present just a very brief reminder.

The general principle to build a land cover map from images is to automatically classify these images. There are several ways to do this but the approach usually adopted is that of supervised learning (use of knowledge *a priori*). For this, it is necessary to define the number and nature of thematic classes in advance. In a landscape ecology approach, this choice must be made in consultation with ecologists to obtain a representation in line with habitats of the species. As stated above, an appropriate spatial resolution and acquisition date should be selected to make the discrimination of the classes possible. Often, two dates are sufficient to obtain a very simple land product (with a date in the winter and the other in the spring or summer to distinguish the deciduous and coniferous trees for example). As soon as one wishes to separate summer and winter crops (or even different types of cultures), it is necessary to select more images on the dates key, referring to the cultural calendar of the study area.

---

5 Download link: http://www.statistiques.developpement-durable.gouv.fr/clc/fichiers/.

6 Download link: http://professionnels.ign.fr/ocsge.

7 Download link: https://www.theia-land.fr/fr/produits/occupation-sols.

We can consider that the construction of a land cover map from images needs to go through the following steps[8]:

– selection and downloading of images;

– potential pretreatments depending on the level of correction of the images;

– mosaic (if necessary, depending on the scope covered by each image);

– sampling of reference data: construction of the training and validation dataset (usually from a survey field but these references can also come from a photointerpretation or auxiliary sources like *Registre Parcellaire Graphique*[9]);

– classification: train the model and projection on the whole of study area;

– validation: calculation of statistical indicators (overall accuracy, kappa coefficient, F-score, rate of deficits and surpluses);

– potential postprocessing: application of a filter of regularization space to remove isolated pixels (majority vote or sieving).

There are many algorithms for supervised learning. Currently, state of the art classifiers are *Support Vector Machines* [MOU 11] or RFs [PEL 16]. They generally provide the best results.

7.3.2. *Step 2: definition of the relevant landscape descriptors*

7.3.2.1. *Choice of indicators*

Characterization of landscapes is driven by performing quantitative analysis based on the calculation of descriptors. These values vary according to:

– the scope of the study site (extent). Often, the "landscape" corresponds to a surface covered by a buffer zone created from a distance that is fixed and centered on the points of biological observations. The distance value depends on the dispersal ability of the species;

– the precision of the cartographic representation (i.e. the level of detail, spatial resolution or "grain");

– the thematic resolution: number of land cover classes chosen according to the species studied (which may require upstream potential groupings of classes).

---

8 A tutorial explaining how to implement a process of classification is available at the following address: https://www.youtube.com/watch?v=L9KCbbJUuCw&t=33s.

9 Geographic data from common agricultural policy CAP available at : https://www.data.gouv.fr /fr/datasets/registre-parcellaire-graphique-rpg-contours-des-ilots-culturaux-et-leur-groupe-de-cultures-majoritaire/.

As mentioned in section 7.2, these descriptors can characterize both the composition of the landscape but also their spatial organization. Moreover, they can be calculated at different levels of analysis (from the patch to the landscape).

Two approaches are possible to select relevant descriptors. The first, exploratory, can be to calculate a very large number of descriptors and then measure their relevance by selection of variables. The second, on the contrary, may limit the number of descriptors to calculate by assumption and relying on ecological theories and knowledge about the species.

7.3.2.2. *Calculation of indicators*

Several tools allow landscape descriptors to be calculated. The oldest is Fragstats software whose use is widespread [MCG 12]. There is also Chloe2012 software that was developed by the INRA[10] of Rennes, France [BOU 14]. A more comprehensive review of these tools can be found in [STE 09].

In this chapter, we use LecoS, a plugin for the QGIS software suite, to calculate various metrics (Table 7.2). The number of metrics available is much smaller than that which is proposed by Fragstats but more conventional metrics can be obtained. LecoS may be applied at landscape scale or on smaller units (a regular grid, polygons of interest). We can import a layer containing these circumscribed areas (landscape vector overlay) before starting the analysis. The landscape map (input data) must correspond to a raster format.

Type of descriptors	Descriptors of the surface, density, shape descriptors, descriptors of diversity.
Levels of analysis	The class (all patches of the same nature) and landscape (all types of patches).
Scope of analysis	Global landscape, irregular or regular grid, area of interest, etc.
Input data	Raster for land cover map – vector for areas of interest.
Output data	Vector layers attribute table or CSV file.
Interests	Free – QGIS plugin – Different format of vector data for areas of interest – Direct mapping of results on QGIS.
Limits	Few descriptors calculated – still in development – no large documentation.

Table 7.2. *Features of metrics calculated under the LecoS plugin. Some computable descriptors (about twenty) describe composition (variety and abundance of the elements) and configuration (spatial characteristics and distribution of the elements) of the landscape. They are inspired by the software Fragstats*

10 *Institut National de la Recherche Agronomique* (French National Institute for Agricultural Research).

7.3.3. *Step 3: statistical modeling*

Statistical analysis that concludes the processing chain aims to assess the effects of the landscape on biodiversity indicators. In our case, we propose simple multivariate modeling, based on a regression model, and in particular, a GLM. Unlike a classical linear model (LM), which makes the assumption that the response variable (or the error term) follows a normal distribution, the GLM makes it possible to take into account other distributions such as a binomial or Poisson law through a link function.

Once the model parameters are estimated, it is then possible to identify the explanatory variables that significantly influence the response variable by referring to the ***P***-value for a fixed risk level α (may reject the hypothesis null Ho -- no significant effect – if it is true). Often, this level of risk is fixed at 5%. In other words, if the ***P***-value < 0.05, the test is deemed significant.

Beyond the effects of each of the variables, we can also evaluate the quality of the fit of the model by calculating the part of the variance explained (or more exactly, in the case of a GLM, the part of the deviance explained, a pseudo-R^2). This gives an indication of the explanatory model performance, but gives no indication of its predictive ability [SHM 10]. The predictive performance can be enjoyed by building a model on part of the sample data and by testing on the remaining part (or cross-validation if the sample size is limited). By comparing observed and predicted values, we can deduce a predictive performance via the estimate of an average error quadratic, for example, or a correlation coefficient between estimated and predicted values.

Finally, when there are many explanatory variables, we can look to reduce their number to limit the complexity of the model (principle of parsimony). A classic solution to achieve this, having eliminated the collinear variables, is to use Akaike Information Criterion (AIC) [AKA 87]:

$$\mathrm{AIC} = -2\ln(\mathrm{L}) + 2\mathrm{k}, \qquad [7.1]$$

where L is the maximized likelihood and k is the number of parameters in the model.

With this criterion (valid when n/k < 40 with n for the number of observations), the deviance of the model $(-2\ln(\mathrm{L}))$ is penalized by the number of variables k introduced. The "best" model is then the one which possesses the lowest AIC provided that the difference (Δ) of AIC > 2. If not (ΔAIC ≤ 2), models are considered to be equivalent.

We can refer to [SHM 10, ZUU 10], or [GUI 00] for more details on the approach to statistical modeling according to this method (including the conditions of application of the models). The practical implementation of this step is performed with R (version 3.1.1) software and its development interface R Studio.

7.4. Application of the processing chain: effect of landscape on forest bird diversity

We offer a practical implementation of the approach on a set of data related to avian diversity. The data available (described in the following sections) are:

– rectangle of the study area (vector layer in projection Lambert 93: "rectangleStudySite_Coteaux.shp");

– points locating counts of birds from 2016 (vector layer in projection Lambert 93: "50PointsCountsBirds_Coteaux.shp");

– land cover map of France from 2016 (raster projection Lambert 93 "OCS_2016_CESBIO.tif" with associated tfw file).

7.4.1. *"Birds" data and the variable to be explained*

Breeding bird communities are well known as being follow-up indicators of landscapes' transformations. Ornithologists from INRA (Dynafor laboratory) studied the evolution of these communities for 40 years in a southwestern French site, near Aurignac village (43° 13'02.76 "N; 0° 52'53.80" E). The objective is to better understand how spatial patterns influence the spatial distribution of species and communities.

To illustrate the method, we will use ornithological surveys on this long-term ecological study. Their biological data date from 2016 and concern forest bird species. Statements were made by adopting the method of the point counts (i.e. *Simple Punctual Sampling* method (EPS)) updated by [SPI 74]. With this method, the birds are identified in their environment through their singing, in spring, during the breeding season. The bird data were taken from 20-min point counts carried out in the month of May between 6 and 11 am, during the peak of vocal activity, in the absence of heavy wind and rain. They noted the presence/absence and abundance of each species using a 250 m point count radius that corresponds to the extent of the home range for most species of passerine birds [RAI 07]. From these surveys, indicators such as species richness of forest birds (i.e. the number of different forest species) were calculated for each point count. In the end, we have a table with points counts in rows (n = 50) and columns, the abundance of each forest species (one column by species) and the derived specific richness.

7.4.2. "Landscape" data and the explanatory variables

We chose to use the land cover map from 2016 of metropolitan France (product OSO) carried out by the Theia center (http://www.theia-land.fr/). This national center aims to make value-added products available to users, in particular, Earth observation data. This map is a raster at 10 m of spatial resolution. The nomenclature consists of 17 thematic classes (Table 7.3, Figure 7.3).

- Summer crop (11) - Winter crop (12) - Deciduous forest (31) - Coniferous forest (32) - Grassland (34) - Mountain pastures (35) - Wood moor (36) - Urban (40, 41) - Diffuse urban (42)	- Industrial and commercial areas (43) - Road (44) - Mineral surfaces (45) - Beaches and dunes (46) - Water (51) - Glaciers or snow (53) - Meadows (211) - Orchard (221) - Vine (222)

Table 7.3. *Nomenclature of the classes of the OSO 2016 map with associated codes. For a color version of the figure, see www.iste.co.uk/baghdadi/qgis3.zip*

Figure 7.3. *The land cover French map OSO 2016 (43°16'23.24" N; 0°48'05.75" E) with an area of 17 km². For the associated nomenclature, see Table 7.3 and http://osr-cesbio.ups-tlse.fr/~oso/ui-ol/S2_2016/layer.html. For a color version of the figure, see www.iste.co.uk/baghdadi/qgis3.zip*

The OSO 2016 product is made completely automatically from data acquired between late 2015 and the end of 2016 with images from the sensor Sentinel-2, and supplemented with images from the sensor Landsat-8[11] [ING 17]. It comes with the

11 See details of the procedure of filing online: http://osr-cesbio.ups-tlse.fr/~oso/.

overall quality statistics: index Kappa[12] = 0.871, and overall accuracy (OA)[13]= 0.889. As we are only interested in the effects of landscape on forest bird species' richness, let us simplify the land cover map to a binary forest/no forest map. For this category, the F-score (average harmonic of recall and precision) is estimated at more than 0.9. From the binary map, the landscape descriptors have been calculated for each circle associated with point counts (buffer zone of radius = 250 m). Six descriptors have been selected in LecoS to measure the heterogeneity of composition (noted COMP) and configuration (CONF) (Table 7.4).

Note that it is also possible to measure the landscape effects without the use of a land cover map but directly using spectral bands of the image or derivatives (such as the *Normalized Difference Vegetation Index*). Such an approach has been adopted in [SHE 14].

Metric	Description	Heterogeneity
Land cover	Area occupied by patches of the same class in each buffer (in map unit)	COMP
Landscape proportion	Proportion occupied by patches of the same class in each buffer (in %)	COMP
NP	Number of patches of the same class.	CONF
LPI	Index of the dominant patch corresponding to the relation between the surface of the largest patch of the class and the total area of the buffer zone. The index is between 0% (larger size, smaller patch) and 100% (landscape consisting of one patch of larger size).	CONF
ENN_MN	Average of the shortest distances to the patch from the same class. The distance is estimated from edge to edge, from the center of the pixel. The value of ENN_MN approaches 0 as the nearest neighbor's distance decreases. Its maximum value has no limit, it is related to the size of the studied site.	CONF
SHDI	Shannon diversity index. This index is 0 when the landscape is composed of a single patch. The value of the index increases with the number of classes of land use or with a regular distribution of surface between the classes.	COMP

Table 7.4. *Six metrics selected for the case study*

12 Kappa: statistical index corresponding to the agreed rate (OA) corrected for the effect of randomness.

13 OA: Overall accuracy or overall precision that gives the total number of correctly classified pixels.

The different metrics' values have very different orders of magnitude and units (%, m², without unit). To visually observe the projection of the sampling points in a graph on these six metrics, we can use the radar chart, taking care to normalize the values of metrics (% of the max value of the scale by factor).

7.4.3. *Implementation in QGIS environment*

The implementation of the approach under QGIS and R software is described below. For color versions of the figures in the following sections, see www.iste.co.uk/baghdadi/qgis3.zip.

Manipulation in QGIS

1. Prepare the input data on our study area

- Create circular buffers of radius 250 m centered on the point counts of birds: *Menu Vector > Geoprocessing Tools > Fixed distance buffer*
  - o Input layer: "50PointsCountsBirds_Coteaux.shp"
  - o Distance: 250
  - o Buffer (output): "Buffer_r250m_50PointsCountsBirds_Coteaux.shp"

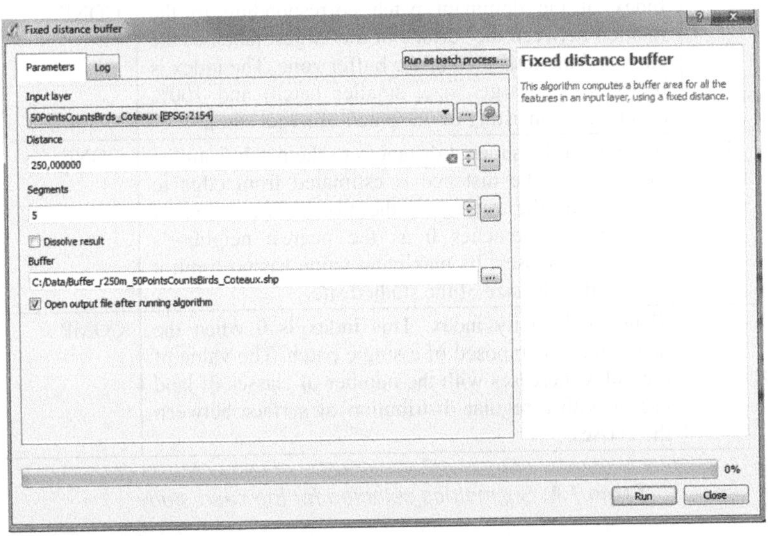

- Clip the land cover map 2016 according to the boundaries of the 107 km² rectangle study site: *Menu Raster > Extraction > Clipper*
  - o Input file (raster) : "OCS_2016_CESBIO.tif";
  - o Output file: "rocs2016.tif";
  - o Mask layer (vector): "rectangleStudySite_Coteaux.shp".

Leave the other default settings. Additionally, it possible to see the *gdal* command line runs from the GUI.

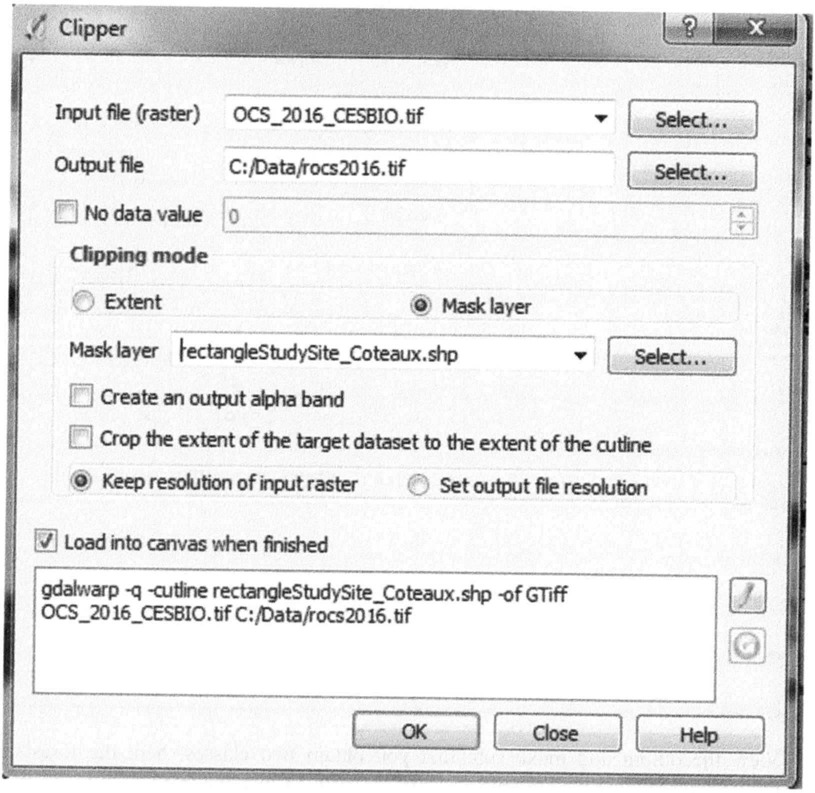

- Reclassify the cropped land cover map into two categories, forest/not forest: *Menu Raster > Raster calculator*
  - Select the layer to be reclassified (raster): "rocs2016.tif";
  - Output layer (raster): "FNFrocs2016.tif";
  - Expression: [("rocs2016@1">= 31 AND "rocs2016@1"<=36) * 1 + ("rocs2016@1" !=31 OR "rocs2016@1" !=32 OR "rocs2016@1" !=34 OR "rocs2016@1" !=35 OR "rocs2016@1" !=36) * 2].

In this way, all pixels whose values are between 31 and 36 will be reclassified in 1 (forest), the other pixels will be set to 2 (not forest).

Other settings can be left as default.

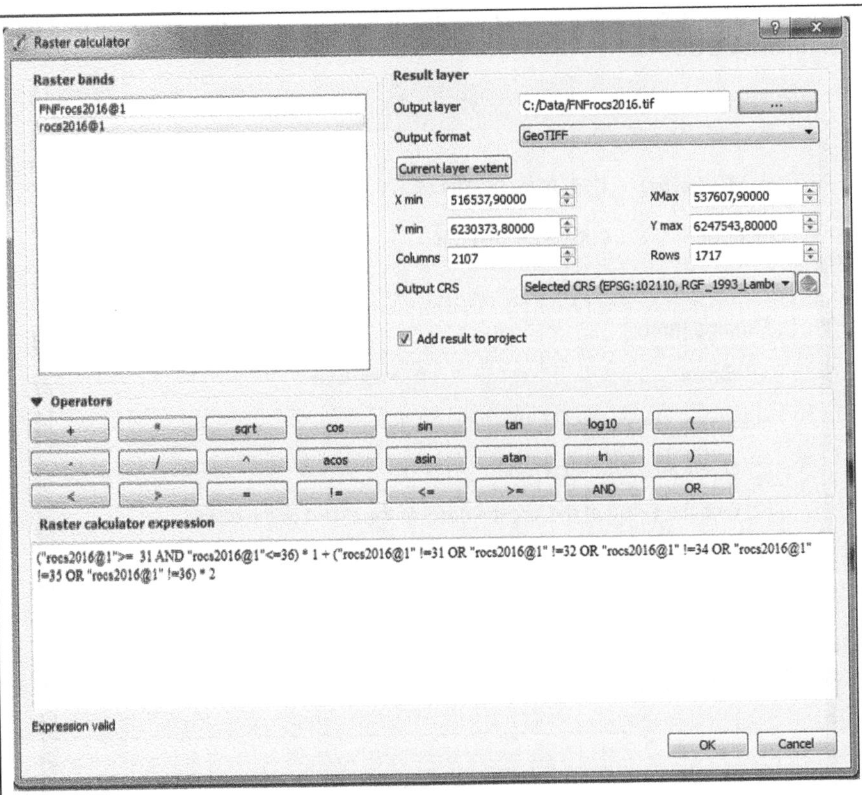

- View the output and make sure that you obtain two classes: here the forests are represented in black and the other classes in white. You can also overlay layers of the point counts of birds and circular buffers of 250 m radius.

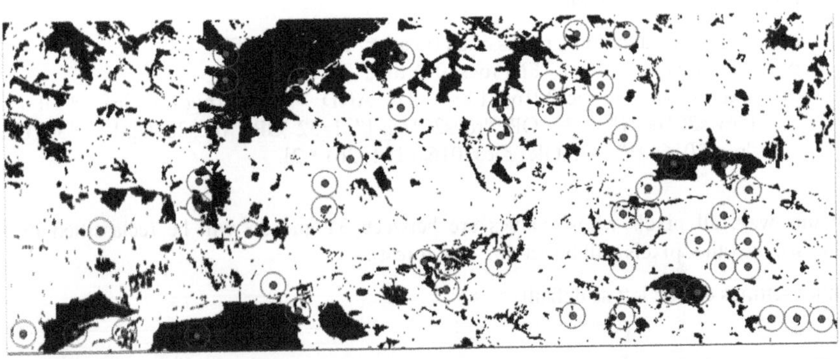

2. Calculate the required landscape descriptors

- Install the LecoS plugin in QGIS: *Menu Plugins > Manage and install plugins*, search LecoS and activate.

- Launch the LecoS module and choose the model: *Menu Raster > Landscape Ecology > Landscape vector overlay*. This module allows us to calculate metrics for our buffers and save the values in the vector layer attribute table.
- In the tool window (1) Polygon overlay batch, enter:
  - Landscape Layer (input): raster "FNFrocs2016.tif";
  - Overlaying grid: "Buffer_r250m_50PointsCountsBirds_Coteaux.shp".
- In the tool window (2), the parameters are the following:
  - Level of analysis: here settings are class and in particular that of forests (class 1).
- In the window (3), the metrics are (see Table 7.4):
  - Forest areas (land cover), proportion occupied by forest patches (landscape proportion LP), number of patches (NP), index patch dominant (LPI), average of the shorter distances to the patch for the same class (ENN) for classes "forest" and "non-forest".
- In the window (4), specify if you want to export values in the layer attribute vector table of "Buffer_r250m_50PointsCountsBirds_Coteaux.shp" in new columns.

After a first pass, restart the tool *Polygon overlay batch* in order to calculate the Shannon diversity index (SHDI) across the landscape (with the same input data).

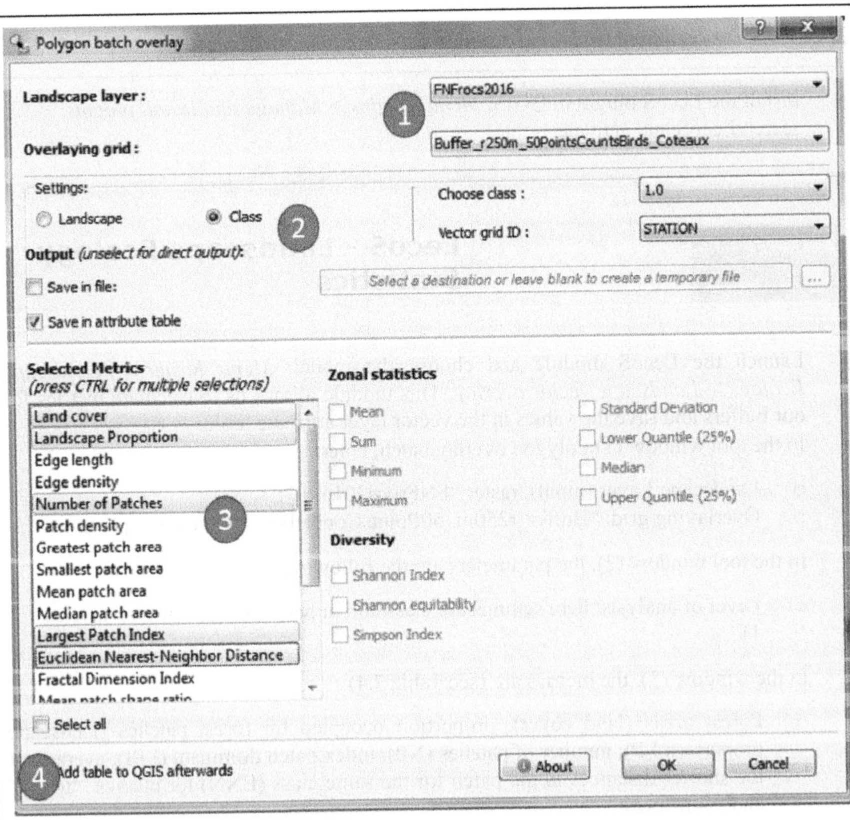

We represent the values of the landscape descriptors (for the class "forest") in a radar chart. The indices were standardized in % of the max value of the scale factors (i.e. reduced to a scale from 0 to 100) to be comparable.

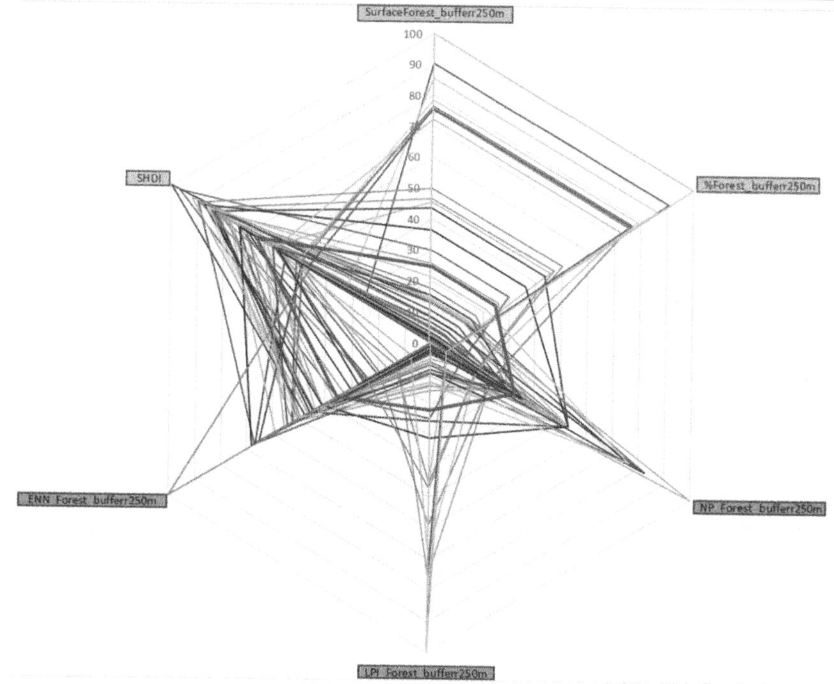

Interpretation:

We notice the overall profile of the 50 point counts on six axes of landscape descriptors, with blue descriptors marking the heterogeneity of composition and violet descriptors marking the heterogeneity of configuration. The average profile is indicated as a red line. The scenery around our sampling points are, on average, diversified with a large gradient of forest amount.

3. Model the birds-landscape relationships in the R software

3.1. *Import library in R software*

Several R packages are needed and must be loaded prior to implementing this study of landscape ecology. These have features to easily reproduce the various stages of our analysis, from the import of the data inputs to the analysis of the effects of landscape variables on the species richness of forest birds. These packages can be installed directly from the CRAN package archive network for the R programming language or through the download of zip files available at the following address: https://cran.r-project.org/package

```
rm(list=ls())# to erase all objects existing in R before running
R code to come
>library(MASS) # to install library MASS
>library(ggplot2) # to install library ggplot2
```

3.2. *Import data tables*

The first step is to import the tables of data in the R software. Two files are available. The "birds_forests.csv" file contains bird data (see section 7.4.1) while the "landscape.csv" contains landscape descriptors. To import these files, we can run the following commands directly in the console of the software.

```
dataO = read.csv("birds.csv", head = TRUE)
dataP = read.csv("landscape.csv", head = TRUE)
```

A few simple commands to obtain the statistics on different data sets:

```
>nrow(dataO) # number of points counts of birds
50
>mean(dataO$Richness) # average of species richness
7.6
>sd(dataO$Richesse) # deviation of species richness
2.5
>sum(dataO$Abundance) # number of individuals identified total
1029
>mean(dataP$SurfaceForest_bufferr250m) # average of forest surface
50000
>sd(dataP$SurfaceForest_bufferr250m) # deviation of forest surface
61084.31
```

A few observations:

A total of 1,029 individuals belonging to 20 species were identified through 50 point counts. Average species richness is 7.6 (±2.5) while the average wooded surface within a radius of 250 m around each point is 50,000 (±61,084) m^2.

3.3. *Withdrawal of collinear variables*

Before the construction of the regression model, please check the correlation between explanatory variables. The introduction of dependent variables can lead to a biased model, sensitive to small variations (high variance). Also, the effect of some variables may not be detected or their relative impact difficult to assess [DOR 13]. To check this dependence, we can calculate the Spearman's rank correlation coefficient (one of the landscape variables does not meet the assumption of normality for the calculation of the Pearson coefficient). If some variables are correlated with more than 70% [DOR 13], then remove the redundant ones. Otherwise ($\rho < 0.7$), they can be stored for later.

The Spearman ρ estimate between variables is described below:

```
# Pearson and Spearman correlation? Verification of the hypothesis of
normality for each of the variables with the test of Shapiro. SHDI
example.
> shapiro.test(dataP$SHDI)

        Shapiro-Wilk normality test

data:  dataP$SHDI
W = 0.94385, p-value = 0.01918

# Calculation of the matrix of correlations between variables using the
Spearman coefficient
> df = data.frame(dataP$SHDI,
                  dataP$SurfaceForest_bufferr250m,
                  dataP$LP_Forest_bufferr250m,
                  dataP$NP_Forest_bufferr250m,
                  dataP$LPI_Forest_bufferr250m,
                  dataP$ENN_Forest_bufferr250m)

> colnames(df) =  c("SHDI","Surface","LP","Nb_Patch","LPI","ENN")
> cor(df, method = « spearman »)
                SHDI     Surface          LP   Nb_Patch         LPI        ENN
SHDI      1.00000000 -0.03471789 -0.03471789  0.6515546 -0.06948379  0.3437913
Surface  -0.03471789  1.00000000  1.00000000 -0.2012701  0.98905162 -0.4264552
LP       -0.03471789  1.00000000  1.00000000 -0.2012701  0.98905162 -0.4264552
Nb_Patch  0.65155456 -0.20127010 -0.20127010  1.0000000 -0.25249370  0.5703133
LPI      -0.06948379  0.98905162  0.98905162 -0.2524937  1.00000000 -0.4278024
ENN       0.34379130 -0.42645518 -0.42645518  0.5703133 -0.42780244  1.0000000
```

Interpretation:

The Shapiro test makes it possible to check the normality of the distribution of the variable passed in parameter (here, SHDI). The hypothesis H_0 of this test is that the distribution follows a normal distribution. Therefore, if the test is significant (*P-value* < 0.05), H_0 is rejected. This means that normality is not respected, which is the case for our variable SHDI (*P-value* = 0.019). For this reason, we select the Spearman's rank correlation.

The matrix of correlations shows that Surface, LP and LPI variables are highly correlated by exceeding the recommended threshold of 0.7. This seems quite logical since on the one hand, we have aggregated classes to get a binary map and on the other, the surface of the patch of the largest forest is close to the total area of forest in the buffer zone. We therefore decide to dismiss the LPI and LP variables to keep only the Surface, the number of patches (NP) and the metric ENN in the analysis.

3.4. *Selection of variables and construction of an explanatory model*

At this stage, we assess the effects of landscape variables selected by the specific richness in forest birds through a GLM. In a regression model, one often seeks to find a compromise between the number of variables to be used and the quality of the fit of the model to the data. Specifically, we are looking for a balance between the bias and the variance by checking the adjustment and the complexity of the model. As mentioned above, the selection of the most parsimonious model can be made according to the AIC criterion full template (including all variables). Therefore, we will proceed with this selection of models before interpreting the effect of each of the variables used in the most parsimonious model. To do this, we will use all of the available data set (case of the construction of an explanatory model). Furthermore, we will transform the Surface variable in log (Surface) because the theoretical area-species relationship follows a power law (logarithmic transformation allows us to have a linear relationship) [CON 79].

```
# To build a table containing the variable response (species richness)
> Y = data.frame(dataO$ID,dataO$Richness)
> colnames(Y) = c("ID","Richness")
# To build a table containing the explanatory variables
> X = data.frame(dataP$ID, dataP$SHDI, dataP$SurfaceForest_bufferr250m, dataP$NP_Forest_bufferr250m, dataP$ENN_ Forest _bufferr250m)
> colnames(X) = c("ID","SHDI","Surface","Nb_Patch","ENN")
> SET = merge(X,Y,by="ID") # Assembly of the 2 tables
# Construction of the full GLM model where we seek to explain the variable Richness based on the independent landscape variables. The distribution of the response variable is assumed to follow a Poisson law (case of a count - link log).
> mod = glm(Richness ~ SHDI + log(Surface) + Nb_Patch + ENN, family = "poisson", data = SET)
# Selection of the most parsimonious model from the model incorporating all the variables (the interpretation is given later)
> stepAIC(mod)
Step:  AIC=215.45
Richness ~ log(Surface)
                 Df  Deviance    AIC
<none>               20.192   215.45
- log(Surface)   1   43.811   237.07
Call:  glm(formula = Richness ~ log(Surface), family = "poisson", data = SET)
Coefficients:
 (Intercept)  log(Surface)
     0.5083        0.1522
Degrees of Freedom: 49 Total (i.e. Null); 48 Residual
Null Deviance:          43.81
Residual Deviance: 20.19        AIC: 215.5

# Construction of the GLM corresponding to the previous result (stepAIC) thus including the variable log(Surface)

> mod_select = glm(formula = Richness ~ log(Surface), family = "poisson", data = SET)
```

```
# You can now get all the parameters of the model:
> summary(mod_select)
Coefficients:
              Estimate  Std. Error  z value  Pr(>|z|)
(Intercept)   0.50829   0.33034     1.539    0.124
log(Surface)  0.15224   0.03203     4.753    2.01e-06 ***
--
Signif. codes:  0 '***' 0.001 '**' 0.01 '*' 0.05 '.' 0.1 ' ' 1
(Dispersion parameter for poisson family taken to be 1)
    Null deviance: 43.811  on 49  degrees of freedom
Residual deviance: 20.192  on 48  degrees of freedom
AIC: 215.45
Number of Fisher Scoring iterations: 4
# Calculation of Pseudo R²
> D2 <- (mod_select$null.deviance - mod_select$deviance) /
mod_select$null.deviance
> D2
[1] 0.5391106
```

Interpretation:

The selection of the most parsimonious model using the AIC criterion led us to retain only the variable Surface. The model Richness ~ log(Surface) is therefore the best compromise between the proportion of variance explained by the model and the number of explanatory variables used (AIC = 215.45). Because we implicitly used a log function type link associated with a Poisson distribution for the response variable, it is not the Richness but more precisely the log(Richness), which is a linear function of log(Surface).

If we analyze the parameters of this model, we can see that the y-intercept (intercept) is worth 0.508 and that the regression coefficient associated with the variable log(Surface) is positive and worth 0.152. In other words, the variable log(Surface) has a positive effect on forest bird richness: the more it increases, the higher richness is. This effect is very significant (*P-value* close to zero).

The *summary()* function also provides information about the deviance of the model (i.e. the gap to perfect fit). A model that perfectly adjusts to the data is a model with as many parameters as observations: it is a saturated model. Conversely, a model which does not include an explanatory variable (if it is not a constant) corresponds to the *null* model. Generally, the gap in terms of log-likelihood (L) between the deviance D measures the gap of log-likelihood (L) between the saturated model (L_{sat}), with a multiplicative factor equal to 2:

$$D = D_{\text{residual}} = 2(L_{\text{sat}} - L)$$

This is what is called residual deviance. *Null deviance* corresponds to the gap in log-likelihood between the saturated model and null model:

$$D_{\text{null}} = 2(L_{\text{sat}} - L_{\text{null}})$$

A model that fits is a model that has a low residual deviance. Here, this residual deviance is worth 20.19 (48 degrees of freedom – 2 parameters) against 43.81 for deviance zero (49 degrees of freedom – 1 parameter). As a reminder, the number of degrees of freedom corresponds to the difference between the number of observations (50 in our case) and the number of parameters. From these indicators, one can calculate a pseudo-R^2 (or D^2) as being written [GUI 00]:

$$D^2 = \frac{D_{\text{null}} - D_{\text{residual}}}{D_{\text{null}}}$$

In our case, it is 0.539. This indicator can be interpreted roughly as the R^2 which is usually used in simple linear regression (LM): the model built from the forest surface accounted for about 54% of the species richness in forest birds.

```
# See the effect of the explanatory variables (i.e. log(Surface)) on
the variable response (richness in forest birds)

> termplot(mod_select, partial.resid=TRUE, se=TRUE,
col.term='black', col.se='black',xlab ="log(Surface)",ylab =
"Partial residues")
```

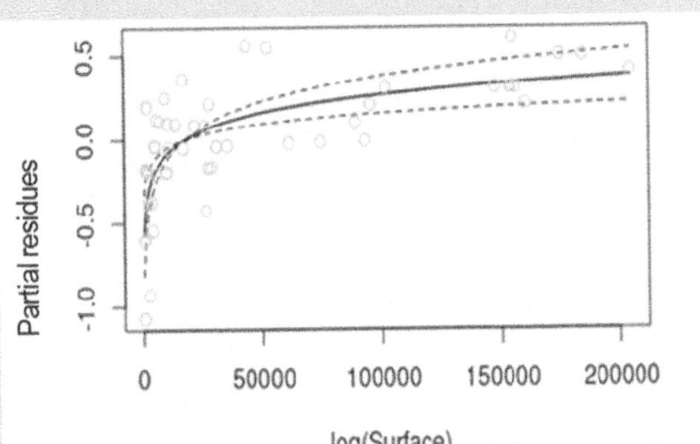

The graph above shows the track of partial residuals (observed value – predicted value) for the variable log(Surface) of the model.

3.5. *To measure the predictive ability of the model*

The last step in our analysis is to estimate the predictive power of the model: what would be the quality of prediction if one planned the estimated model on a new data set? For this, we will not use all the observations to build the model (case of the explanatory model) but only a part of the observations in order to test the model on an independent sample. Given the limited number of observations, we will proceed by cross-validation, using the *leave-one-out* method. In this procedure, a single observation is removed for the test (validation) while the other observations are used to estimate the model. This (withdrawal) selection is repeated as many times as there are observations (50 in our example). Predictive performance is then estimated here by calculating the average of errors, or the average difference between the observed and predicted values. We could also calculate the average quadratic deviation or the correlation between predicted and observed.

```
# Creation of a vector to accommodate the predicted values

> Richness_predicted = vector(mode="numeric", nrow(SET))

# Construction dataset of calibration and test
# Estimation of a linear model from the calibration set
# Projection of the model on the test dataset

> for (i in 1:nrow(SET)){

+   calib=SET[-i,]

+   eval=SET[i,]
+   mod_calib = glm(Richness~log(Surface),data = calib, family =
"poisson")

+   R_pred = predict(mod_calib,eval, type = "response")+   Richness_
predicted [i] = R_pred}

> Richness_observed = SET$Richness

error_mean = > mean(abs(Richness_ predicted -Richness_observed))

[1] 1.385949

# Scatter plot between predicted richness and observed richness

> tablePlot = data.frame(Richness_observed, Richness_predicted)

> qplot(Richness_observed,Richness_predicted, data = tablePlot)
+geom_smooth(method=lm,se=FALSE,color = 'dimgray',size = 0.5)
```

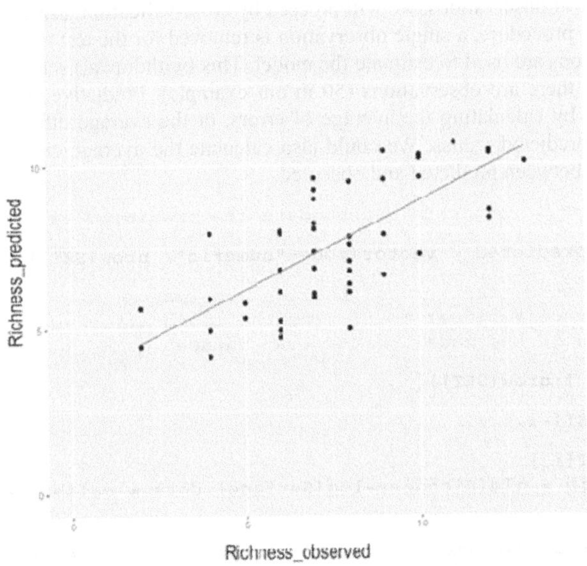

The average prediction error is equal to 1.39 (so a little more than one forest bird species).

7.5. Acknowledgments

The authors wish to thank Gérard Balent, Research Director of Ornithological Research, and Laurent Raison, technician ornithologist, at the Dynafor Laboratory, for having given their agreement to use a sample of avifauna data in this study.

7.6. Bibliography

[AKA 87] AKAIKE H., "Factor analysis and AIC", *Psychometrika*, vol. 52, pp 317–332, 1987.

[BOU 14] BOUSSARD H., BAUDRY J., Chloe212: a software for landscape pattern analysis, available at: http://www.rennes.inra.fr/sad/Outils-Produits/Outils-informatiques/Chloe, 2014.

[CHA 76] CHATFIELD C., GOODHARDT GJ., "The beta-binomial model for consumer purchasing behavior", in FUNKE U. (ed.), *Mathematical Models in Marketing*, Springer, Berlin, Heidelberg, 1976.

[CON 79] CONNOR EF., MCCOY ED., "The statistics and biology of the species-area relationship", *American Naturalist*, vol. 113, no. 6, pp. 791–833, 1979.

[CUS 08] CUSHMAN S., MCGARIGAL K., NEEL M., "Parsimony in landscape metrics: strength, universality, and consistency", *Ecological Indicators*, vol. 8, no. 5, pp. 691–703, 2008.

[DEC 04] DECAMPS H., DECAMPS O., *Au printemps des Paysages*, Buchet-Chastel, Paris, 2004.

[DOR 13] DORMANN F.C., ELITH J., BACHER S. *et al.*, "Collinearity: a review of methods to deal with it and a simulation study evaluating their performance", *Ecography*, vol. 36, pp. 27–46, 2013.

[DUE 97] DUELLI P., "Biodiversity evaluation in agricultural landscapes: an approach at two different scales", *Agriculture, Ecosystems and Environment*, vol. 62, pp. 81–91, 1997.

[FAH 11] FAHRIG L., BAUDRY J., BROTONS L. *et al.*, "Functional landscape heterogeneity and animal biodiversity in agricultural landscapes", *Ecology Letters*, vol. 14, pp. 101–112, 2011.

[FOR 86] FORMAN R.T.T., GODRON M., *Landscape ecology*, Wiley, New York, 1986.

[GUI 00] GUISAN A., ZIMMERMANN N.E., "Predictive habitat distribution models in ecology", *Ecological Modelling*, vol. 135, pp. 147–186, 2000.

[GUS 98] GUSTAFSON E.J., "Quantifying landscape spatial pattern: what is the state of the art?", *Ecosystems*", vol. 1, pp. 143–156, 1998.

[ING 16] INGLADA J., "Land Cover Mapping from Optical Images", in BAGHDADI N., ZRIBI M., (eds), *Land Surfaces Remote Sensing in Agriculture and Forest*, ISTE Press Ltd, London and Elsevier Ltd, Oxford, 2016.

[ING 17] INGLADA J., VINCENT A., ARIAS M. *et al.*, "Operational high resolution land cover map production at the country scale using satellite image time series", *Remote Sensing*, vol. 9, no. 95, 2017.

[LAN 05] LANCELOT R, LESNOFF M., Sélection de modèles avec l'AIC et critères d'information dérivés, available at ftp://ftp.cirad.fr/pub/group-r/groupe-r/Fiches/AIC_v3.pdf, 2005.

[LE 13] LE REST K., Méthodes statistiques pour la modélisation des facteurs influençant la distribution et l'abondance de populations : Application aux rapaces diurnes nichant en France, PhD thesis, University of Poitiers, 2013.

[MAC 67] MACARTHUR R.H., WILSON E.O., *The Theory of Island Biogeography*, Princeton University Press, Princeton, 1967.

[MCG 12] MCGARIGAL K., CUSHMAN S.A., ENE E., FRAGSTATS V4: Spatial Pattern Analysis Program for Categorical and Continuous Maps, University of Massachusetts, Amherst, available at : http://www.umass.edu/landeco/research/fragstats/fragstats.html, 2012.

[MOI 16] MOISJA K., UUEMAA E., OJA T., "Integrating small-scale landscape elements into land use/cover: the impact on landscape metrics' values", *Ecological Indicators*, vol. 67, pp. 714–722, 2016.

[MOU 11] MOUNTRAKIS G., IM J., OGOLE C., "Support vector machines in remote sensing: a review", *ISPRS Journal of Photogrammetry and Remote Sensing*, vol. 66, pp. 247–259, 2011.

[PEL 16] PELLETIER C., VALERO S., INGLADA J. *et al.*, "Assessing the robustness of random forests to map land cover with high resolution satellite image time series over large areas", *Remote Sensing of Environment*, vol. 187, pp. 156–168, 2016.

[RAI 07] RAISON L., "L'inventaire des oiseaux nicheurs par la méthode des points d'écoute dans les paysages agricoles", *Cahier des Techniques de l'INRA*, http://prodinra.inra.fr/record /22391, pp. 79–86, 2007.

[RII 95] RIITTERS K.H., O'NEILL R.V., HUNSAKER C.T. *et al.*, "A factor analysis of landscape pattern and structure metrics", *Landscape Ecology*, vol. 10, pp. 23–40, 1995.

[SHE 14] SHEEREN D., BONTHOUX S., BALENT G., "Modeling bird communities using unclassified remote sensing imagery: effects of the spatial resolution and data period", *Ecological Indicators*, vol. 43, pp. 69–82, 2014.

[SHM 10] SHMUELI G., "To explain or to predict?", *Statistical Science*, vol. 25, pp. 289–310, 2010.

[SIR 16] SIRAMI C., "Biodiversity in heterogeneous and dynamic landscapes", *Oxford Research Encyclopedia of Environmental Science*, 2016.

[SIR 17] SIRAMI C., DECONCHAT M., Nouveaux horizons pour l'Ecologie des paysages, Regards RO4 de la Société Française d'Ecologie, available at https://www.sfecologie.org/regard/ro4-ecologie-des-paysages-m-deconchat-et-c-sirami/, 2017.

[SPI 74] SPITZ F., Facteurs de répartition de l'avifaune en forêt de montagne. Ecologie forestière, la forêt: son climat, son sol, ses arbres, sa faune, Pesson éditeur, Gauthier-Villars, Paris, pp. 327–334, 1974.

[STE 09] STEINIGER S., HAY G.J., "Free and open source geographic information tools for landscape ecology", *Ecological Informatics*, vol. 4, pp. 183–195, 2009.

[TUR 01] TURNER M.G., GARDNER R.H., O'NEILL R.V., *Landscape ecology in theory and practice: pattern and process*, Springer Verlag, New York, 2001.

[UUE 09] UUEMAA E., ANTROP M., ROOSAARE J. *et al.*, "Landscape metrics and indices: an overview of their use in landscape research", *Living Reviews In Landscape Research*, vol. 3, no. 1. 2009.

[WED 74] WEDDERBURN R.W., Quasi-likelihood functions, generalized linear models, and the Gauss–Newton method, *Biometrika*, vol. 61, pp 439–447, 1974.

[WEG 16] WEGMANN M., LEUTNER B., DECH S., *Remote Sensing and GIS for Ecologists: Using Open Source Softwares*, Pelagic Publishing, Exeter, 2016.

[ZUU 09] ZUUR A., IENO E.N., WALKER N. *et al.*, *Mixed Effects Models and Extensions in Ecology with R*, Springer-Verlag, New York, pp. 574, 2009.

[ZUU 10] ZUUR A., IENO E.N., ELPHICK C.S., "A protocol for data exploration to avoid common statistical problems", *Methods in Ecology and Evolution*, vol. 1, pp. 3–14, 2010.

List of Authors

Rémi ANDREOLI
BLUECHAM SAS
Nouméa
France

Josselin AVAL
ONERA
DOTA
Toulouse
France

Nicolas BAGHDADI
Irstea/University of Montpellier
TETIS
Montpellier
France

Christina CORBANE
European Commission
Joint Research Centre
Space, Security and Migration
Disaster Risk Management (JRC.E.1)
Ispra
Italy

Julita DUDEK
CNRS / UBFC
CRC - Biogéosciences
Dijon
France

Justin EMERY
CNRS / UBFC
ThéMA / CRC - Biogéosciences
Dijon
France

Thierry ERUDEL
ONERA
DOTA
Toulouse
France

Mathieu FAUVEL
INP-ENSAT
DYNAFOR
Toulouse
France

Ludovic GRANJON
CNRS / UBFC
MSH de Dijon, GeoBFC
Dijon
France

Pierre-Alexis HERRAULT
Strasbourg University
CESBIO
Toulouse
France

Thomas KEMPER
European Commission
Joint Research Centre
Space, Security and Migration
Disaster Risk Management (JRC.E.1)
Ispra
Italy

Sylvie LADET
INRA
DYNAFOR
Toulouse
France

Clément MALLET
IGN-ENSG/ University of Paris-Est
aSTIG/MATIS
Saint-Mandé
Paris
France

Nadège MARTINY
CNRS / UBFC
CRC - Biogéosciences
Dijon
France

Boris MERICSKAY
University of Rennes 2
Espaces et Sociétés
Rennes
France

Martino PESARESI
European Commission
Joint Research Centre
Space, Security and Migration
Disaster Risk Management (JRC.E.1)
Ispra
Italy

Benjamin POHL
CNRS / UBFC
CRC - Biogéosciences
Dijon
France

Panagiotis POLITIS
Arhs Developments S.A.
Luxembourg

Yves RICHARD
CNRS / UBFC
CRC - Biogéosciences
Dijon
France

David SHEEREN
INP-ENSAT
DYNAFOR
Toulouse
France

Alice SIRAGUSA
European Commission
Joint Research Centre
Growth and Innovation
Territorial Development Unit
Ispra
Italy

Thomas THEVENIN
CNRS / UBFC
ThéMA
Dijon
France

Brice Van HAAREN
WWF
New Caledonia

Mehrez ZRIBI
CNRS
CESBIO
Toulouse
France

Index

A, B

C

D, E

F, G

Scientific Committee

Other titles from

iSTE

in

Earth Systems – Environmental Sciences

2018

BAGHDADI Nicolas, MALLET Clément, ZRIBI Mehrez
QGIS in Remote Sensing Set
Volume 1 – QGIS and Generic Tools
Volume 2 – QGIS and Applications in Agriculture and Forest

2017

QUEVAUVILLER Philippe, CIAVOLA Paolo, GARNIER Emmanuel
Management of the Effects of Coastal Storms: Policy, Scientific and Historical Perspectives

2016

BLAIN Stéphane, TAGLIABUE Alessandro
Iron Cycle in Oceans

BRAUDEAU Erik, ASSI Amjad, MOHTAR Rabi
Hydrostructural Pedology

MONACO André, PROUZET Patrick
Seas and Oceans Set
Volume 9 – Tools for Oceanography and Ecosystemic Modeling

QUÉGUINER Bernard
The Biogeochemical Cycle of Silicon in the Ocean

QUEVAUVILLER Philippe
Marine Chemical Monitoring

2015

HERBIN Hervé, DUBUISSON Philippe
Infrared Observation of Earth's Atmosphere

MERCIER-LAURENT Eunika
The Innovation Biosphere

MONACO André, PROUZET Patrick
Seas and Oceans Set
Volume 7 – Marine Ecosystems: Diversity and Functions
Volume 8 – Governance of Seas and Oceans

2014

DE LARMINAT Philippe
Climate Change: Identification and Projections

MONACO André, PROUZET Patrick
Seas and Oceans Set
Volume 1 – Ocean in the Earth System
Volume 2 – The Land-Sea Interactions
Volume 3 – Ecosystem Sustainability and Global Change
Volume 4 – Vulnerability of Coastal Ecosystems and Adaptation
Volume 5 – Development of Marine Resources
Volume 6 – Value and Economy of Marine Resources

MUTTIN Frédéric
Marine Coastal and Water Pollutions: Oil Spill Studies

2013

LÉCUYER Christophe
Water on Earth: Physicochemical and Biological Properties

LEGCHENKO Anatoly
Magnetic Resonance Imaging for Groundwater

2012

LE MENN Marc
Instrumentation and Metrology in Oceanography

PAVÉ Alain
Modeling of Living Systems: From Cell to Ecosystem

2010

TANGUY Jean-Michel
Environmental Hydraulics Series
Volume 1 – Physical Processes and Measurement Devices
Volume 2 – Mathematical Models
Volume 3 – Numerical Methods
Volume 4 – Practical Applications in Engineering
Volume 5 – Modeling Software

Printed and bound by CPI Group (UK) Ltd, Croydon, CR0 4YY
16/09/2026
14963075-0005